Honey show classes

This book is written for Carole and Amy

I would like to thank the following people who have helped with this book either by encouragement or permission to use photographs:
Hazel Blackburn
Enid Brown
Gerry Collins
Terry Ashley
Bernard Diaper, to whom I owe special thanks for working tirelessly developing the BD honey grading glasses, without which we would not be showing or judging honey.
Frank Kirkham
Joyce Nisbet
Charles Davies
Moyra Davidson

Honey show classes
A guide for competitors, organisers and judges

John Goodwin

Honey show classes: a guide for competitors, organisers and judges

© John Goodwin
ISBN: 978-1-914934-17-9

Second edition published in 2022 by:
Northern Bee Books, Scout Bottom Farm, Mytholmroyd, Hebden Bridge,
HX7 5JS (UK). Tel: 01422 882751. www.northernbeebooks.co.uk
Book design: www.SiPat.co.uk

Contents

Foreword 11

1. Duties of a show secretary 13
Show venue 13
    Criteria for booking a venue 14
Planning 18
    Equipment/kit for the day 20
    Method of allocating entries 21
    Layout of the classes 23
    Procedures for the show day 24
    Virtual honey shows 26

2. Schedules 27
Extracted honey 28
    Commercial classes 29
    Black jar class 29
Comb honey 29
    Frames for rotary extraction 29
    Cut comb 29
    Sections 30
Wax 30
    Beeswax blocks 30
    28-Gram blocks of beeswax 31
    Beeswax candles 32
    Beeswax flowers, fruit and models 33
    Beeswax wraps 34

Displays 34
Confectionery 35
    Honey cakes 35
    Honey biscuits 36
Bee forage (flowers) 37
Pot plant 38
Children's class 38
Cosmetics 38
Invention 38
Products of the hive 39
Honey beer 39
Honey vinegar 39
Mead 39
Microscope slides 40
Photographs 41
Videos 42
Honey label 42
Essays 43
Handicrafts 43
Gift classes 44
Observation hives 44
Nucleus hives 45
Skeps 46

3. Stewarding 47
Judge's steward 47
    Kit required 49

4. The secrets of showing 49
Extracted honey 49
    Lids 51
    Labelling 53
    Honey classification by colour 53
    Production 55
    Extraction 55
    Show preparation 56
Soft set honey 57
    Kit required 57
    The process 58
Granulated honey 60
Black jar class 60
Commercial class 61
Chunk honey 63
    Showing 64
    Storage 64

Gift class	64
Composite class	65
Heather honey	66
Extracted heather honey	67
Heather section	68
Heather frame for extraction	69
Comb honey	69
Frame for extraction	69
Cut comb	71
Sections	73
Wax	76
Single wax block	76
The mould	77
Showing	80
Storage	81
28-Gram blocks of beeswax	81
Mould treatment	81
Moulding	82
Showing	82
Storage	82
Commercial wax block	83
Storage	84
Candles	84
Poured	86
Rolled	87
Moulding	87
Fettling and polishing your candles	88
Silicone moulds and other soft plastic moulds	89
Dipping candles: the process	89
Wax models/flowers	90
Beeswax polish	91
Display classes	92
Victorian pyramid	92
Method of showing	93
Shop window display	93
Mead	94
Type of mead	94
Acids	95
Water	96
Honey	96
Yeast	96
Tannin	97
Vitamins and supplements	97
Cold pre-fermentation process	98
Hot water process	99

Honey beer 99
    Kit required 99
Confectionery 100
    Showing 100
Hives: Observation hives 101
    Setting up your observation hive 101
Hives: Nucleus hives 104
Handicrafts 105
Honey jar label and other design classes 105
Essays 105
Videos 106

5. Judging 107
Judges' tool kit and personal protective equipment (PPE) 108
    Instrument specification notes 109
Honey 110
    Judging of extracted runny honey 110
        Kit required 110
        Method 110
Judging soft set and granulated honey 112
    Kit required 112
    Method 112
Judging heather honey 114
    Kit required 114
    Method 114
Judging the black jar class 116
    Kit required 116
    Method 116
Judging the composite class 116
    Kit required 116
    Method 117
Wax 119
Judging single block 119
    Kit required 119
    Method 119
Judging small wax blocks 28g 120
    Kit required 120
    Method 120
Judging the commercial wax block 121
    Kit required 121
    Method 121
Judging candles: various manufacturing methods 122
    Kit required 122
    Method 122
Wax models and flowers 125

Mead	126
Kit required	126
Method	127
Judging honey beer	130
Kit required	130
Method	130
Judging confectionery	132
Honey cake or honey fruit cake	132
Kit required	132
Method	132
Hives	134
Judging the observation hive	134
Kit required	134
Method	134
Judging the nucleus hive	136
Kit required	136
Method	136
Judging a display	138
Judging skeps	139
Kit required	139
Judging method that could be used	140
Handicrafts	140
Kit required	140
Method	140
Children's class	141
Microscope slides	141
Kit required	141
Photographs	141
Kit required	141
Method	142
Videos	142
Open judging	143
Final speech	144
About the author	145

Foreword

This book has been written as a guide for everyone involved with honey shows, be it organising, exhibiting or judging. If on the day of the show the book has helped to achieve successful organisation, high standards of exhibiting, well-written and unambiguous honey show schedules, increased interest in honey shows, and more applicants for the BBKA Show Judge certificate, the writing of the book will have been worthwhile.

The chapter on show management is intended as a guide both for large shows and for minor, branch association shows. What is not understood is that larger shows are very often easier and require less work, since County, Floral and Agricultural shows have the resources, facilities and show management that make organisation somewhat simpler for the Honey Section show manager. In contrast the show manager of small shows may be responsible for all the show organisation required, both prior to the show and during the showing period. Hopefully the reader will find in this chapter a simplified but comprehensive guide of show management.

When competing in honey shows you should always enter to win. This book explains what is expected of the exhibit in each class and category. The importance of the boring exhibit containers (honey jars, boxes and bags) is stressed, in addition to their contents, the equipment, and the required method for showing and producing winning exhibits for each class and category.

Honey show judging is necessarily evolving with the requirements of modern day-to-day beekeeping. Examples include the exhibition of microscope slides, videos and on-line essays. The judging section is intended for prospective judges starting their exam portfolio or well on the path to becoming qualified Honey Show Judges. For all the categories there is a list of the equipment the prospective judge should have to hand, an advised method and procedure and the responsibilities and pitfalls of which a prospective honey judge should be aware.

No apology is given for repetition of the equipment required to judge the various categories.

1
Duties of a show secretary

Running a honey show is no different from any other project with a critical path. The responsibility varies depending upon the function's size, custom and rules, association, society and show committee.

Show venue

If responsible for the venue, the show secretary should book at the earliest moment. If it is a major society or association show, and the honey show is just a section of a larger exhibition, obviously there is no need to book the venue. However, it is necessary to fight your corner for the size and position of the honey section within the show. Show managers sometimes have to allow space for demonstrations and trade stalls, be it candle-making, videos of various aspects of beekeeping, studies of other species of bees and insects, honey bee products and various trade suppliers. All these aspects need space, and this space requirement needs to be considered prior to the booking of the show venue or bidding for space allocation at a larger show.

If the venue is not already decided, this is the first item of the project to be organised. Very often as soon as one show is finished, the date for the next year's show is booked and the venue booking for next year is confirmed.

Secondly, get your judges booked at this moment in time – there are hardly enough BBKA national honey judges to cover the honey shows normally scheduled during the year throughout the UK, especially in late summer and autumn which is the peak period for honey shows.

A good guideline for the number of judges required is one judge per 120 entries.

If possible, please allow a potential judge who is working through their exam portfolio to have a chance of judging your show. If the show has fewer than 100

Honey show classes: a guide for competitors, organisers and judges

1. Tiered staging shelves (Gerry Collins)

entries, this is ideal for a student judge. However, if the show requires more than one judge due to the number of entries, a student judge could be included in your prospective judging team. A second request, perhaps not for the same show year: if you have a request to allow a judge to carry out a student judge assessment, please agree. It may take the judge a little longer to judge your show, but look on it as an investment for the future of honey showing.

Just a note or reminder, you will need a qualified National Honey Show judge to issue a blue ribbon for Best in Show, if you are expecting more than 100 exhibits for your show.

Criteria for booking a venue

This is a tick list you may find useful:
- The most important check, and the first that should be made, is the suitability of the venue, with adequate access and facilities including parking for disabled people, whether members of the public, show staff (stewards), exhibitors or other beekeepers.
- Space: how many exhibits? The calculation is: area of exhibit less the number of tiered staging shelves. (See photo 1).
- Provision of chairs, and tables (covered with a white cloth or disposable paper covering). It is a good idea to use paper if possible, thus saving on washing and ironing white sheets.
- If you are lucky enough to have funds and resources to use a conference centre or an hotel's hospitality suite, a lot of infrastructure work is included in the rental price. Tables are normally clothed and placed to suit your

Duties of a show secretary

2. Observation hives (Enid Brown)

requirements. It is always a good plan to make sure that the hotel or conference centre, fully understands these.
- Provision of chairs for stewards, tellers and managers. These items sometimes have to be rented so an estimate of the requirement is needed early on in the project.
- Lighting: the venue requires to be well lit to enable effective judging of all the exhibits.
- Candles have to be lit to enable judging. If you have candle classes, naked flames will be needed during judging. (See further notes on judging of various exhibit categories). Check with the landlord of the venue if they will authorise you to have naked flames on the premises and that they are insured for this, or that your association is insured in the case of fire. Check the state of play if there are smoke alarms – it would be terribly embarrassing to have the fire service automatically called out to your show!
- Do not expect to remove the candles to a van or Land Rover for judging. All this does is to increase the chance of an accident through movement.
- Judging of mead will require the landlord of the venue to agree to the consumption of alcohol on the premises: this is a very important check if you intend to use a hall or premises used for worship. Some religious denominations are anti consumption of alcohol on their premises. It is not good enough to expect the judge to adjudicate on mead in the car park.
- Make sure whether or not the landlord is happy to have free-flying bees, depending upon the show classes (nucleus and observation hives). (See photo 2).
- Risk assessment: e.g. an observation hive at a public event. Please note this is an example only. All shows should write their own risk assessments.

The nature of the risk	What is done to reduce the risk to a minimum
Breakage of glass.	Hive is securely strapped to avoid falling over. Public are under observation at all times. People are prevented from knocking the glass.
Escape of bees.	All joints, especially exit tube, are secured. Public prevented from moving hive in any way – i.e., securely strapped.
Flying bees from tube bother members of the public.	Tube exit must be positioned so that bees fly out above head-height and in a direction away from the public and towards bushes etc. Tube must be securely fastened at all points – particularly joints.
Deliberate vandalism of hive.	Observation Hive (OH), supervised by at least one person whenever public is allowed into area. OH, securely fastened, and with safety accessories as outlined above.
Bees overheat if hot day.	Water spray available at all times, and person in charge instructed in how to use it. OH, set up out of direct sunlight.
Member of public gets stung.	Avoid any contact between bees and public by measures outlined above.
Observation hive falls over or is pulled over.	OH, securely strapped as above.

Duties of a show secretary

What actions will be prepared for if the risk materialised	Preparations necessary
If glass just cracked – apply wide packing tape. If break severe and danger of bees escaping, apply wood cover to that side. If bees begin to escape or it is likely, cover Observation Hive (OH) with cloth and strap cloth to hive structure.	All the following stored below the OH table: wide packing tape, wood covers, sheet which covers the OH completely. Straps which securely hold OH to table. Extra strapping in case of accident. Spare bee suit. Instruction sheet with contact telephone numbers.
Escaped bees killed if necessary, either by spraying (with soapy water) or swatting. Do not kill bees with an audience – keep public away if necessary. Adhesive tape used to seal up any holes. See above for wooden covers and sheet.	Fly swat, soapy water spray (clearly labelled as such), wood covers, wide packing tape & sheet as above. Strapping to secure the sheet.
Keep public away from tube exit. If necessary, apply bee escape so that no more bees can exit. Do not block tube exit as this will mean returning bees are unable to enter and cause more problems.	One-way bee escape provided as a gate that can be set in an emergency in order to prevent further bees exiting.
Apply security measures above if damage has been done or bees escape. Remove culprit from area and report to show staff.	One person on duty at all times during opening hours. See above for other equipment. Hive to be removed during closing times.
Give bees some water if necessary.	Water spray (clearly labelled to distinguish it from soapy water spray).
Make sure sting immediately removed. Move person calmly away from bees. If they have any medication for stings, advise to take it. Do not administer any medication. Check for allergic reaction. If this is severe, call ambulance immediately. Sit down if possible, or advise to go and sit in the Red Cross tent.	Use mobile phone firstly to call emergency services. Secondly call show organiser. All contact numbers available on briefing sheets supplied to the stewards.
Move public away. If deliberate, make sure culprit is removed from area. Re-erect OH if possible.	Spare strap. Covers as above.

Volunteers and demonstrators should read the risk assessment and sign below

....

- If the honey show is just a section of a larger show (such as floral or agricultural), be aware that there will be procedures in place for evacuation in the event of an imminent disaster (e.g. bomb threat). It is prudent to make all of your team aware of what procedures and evacuation plans are in place, so that in the event of a disaster there is someone on the stand to take responsibility and manage the evacuation as required.
- Heating, ventilation: if it is a summer venue, air conditioning is ideal. (See note on judging soft set honey).
- Facilities: toilets, personal washing facilities. Let's not forget we are promoting and showing honey, which is a food product.
- Catering can be a simple kitchen capable of providing hot and cold water, or have the equipment and space for providing a complete meal, depending upon the needs and expectations of all the show stakeholders.
- Do the landlords allow outside catering or self-catering? Some premises insist on their own catering supply.
- Car park: check for charges for parking.
- Disabled facilities: access and car park allocation.
- Ease of getting to, difficulty in finding, and relevance of location to the geographical catchment area of the show. Make sure the venue's satnav matches the postcode.

Planning

The show day's agenda and day's organisation should be discussed and agreed with the relevant committee. A typical agenda at the above meeting would be:
- Venue.
- Date and time.
- As already stated, if you expect above 100 entries and if it is intended to issue a National Honey Show Blue Ribbon, it will require a qualified National Honey Show judge.
- Last date for entries. Set up deadline for entries, normally allowing at least a week for the return of class entry slips. Never get into the situation where entries are accepted on the day of the show.
- Schedule classes. (It is worth asking members of the show committee showing at or visiting other shows to look out for ideas for different classes).
- Sponsors are worth pursuing, not just for the money sponsorship of classes but also to raise the profile of honey shows both nationally and locally. A good number of companies value the chance to sponsor honey shows, not only to market their products but also to inform their prospective and regular customers about their stakeholding in the local and wider environment.

Duties of a show secretary

Remember that honey bees have fantastic environmental credibility.

- Prize money: for guidance, this is normally just a token amount which does not cover the cost of showing. In the golden years of produce shows, prize money value was equivalent to a week's wage for a typical farm worker.
- If prize money is offered, make sure there is a full audit trail for the money and all money given is signed for.

3. Prize cards (John Goodwin)

- Prize cards: if the show has a valued reputation for its quality of exhibits and competitive merit, and is deemed highly regarded among exhibitors, prize money becomes irrelevant, but the quality of prize cards has to be excellent (photo 3). Just as an aside, it should be the aim of all show managers to achieve the best quality competitive show in the country. With a good novice section, and with novelty classes keeping up with the needs of modern beekeeping, the show will be inclusive for all beekeepers and will keep increasing the number of exhibits, thus having a good reputation for years to come.
- If it is decided to adapt the honey show rules to other than those of the BBKA or the National Honey Show, they should be printed onto the schedule. A typical example would be Commercial 10-12 jar classes using other than 1lb squat jars, and wax polish as for sale.
- Always have a copy of the rules and schedule available in case there is a dispute.
- In case of any dispute, have a stated procedure and a method of resolution.
- Trophy return: it is good practice to arrange prior to the show a letter or email sent to all of last year's winners.
- Engraving: arrange to have an on-site engraver, or for trophies to be presented after the show but returned for engraving and reissued to the winners later. Keep a list of the previous year's winners.
- Manning prior to and during the show
- Demonstrations
- Commercial trade stands

A typical team to help the Show Secretary could be:
- Booking-in clerk
- Trophy manager (responsible for safe return, polishing and engraving)
- Staging team
- Parking attendant
- Staging stewards
- Judges' stewards
- Prize money stewards
- A gifted calligrapher for the show cards
- Judges
- Judges' referee

Just a note on trophies: many trophies have some age and a long history. As show manager, it is worth going to the trouble of documenting what is known of each trophy. One of the oldest known is the Wrekin Perpetual Challenge Vase given at Shrewsbury Flower Show.

Equipment/kit for the day

- Torch for use by judge in case of failure of his or her own.
- Scales.
- Test weight.
- Candleholders if specified by schedule for class.
- Matches.
- Hammer if there is a commercial class for wax.
- Microscope if there is a class for mounted slides; and also a method of exhibiting the slides rather than have the public using a valuable microscope.
- Comment slips for the judges.
- Judging sheets: see examples following.
- Supply of white coats, hats and disposable gloves. If the convention is that stewards provide their own, make sure that they are aware of the requirement to arrive at the venue with the necessary kit.
- Disposable plastic gloves.
- Honey jar legislation (download).
- BBKA list of plants that honey bees will work or benefit from.
- If there are classes for hives – both observation and nucleus – have available all the necessary PPE, including bee suits, hive tools, gaffer tape, emergency dust sheets.
- If there is a class for nucleus hives, a cage will have to be provided.
- Honey grading glasses must now be BD 2014 grade: purchase from Thorne's (photo 4).
- ISO wine glasses (judges normally provide their own).
- Plastic domes for showing cakes, confectionery and pieces of blocks of beeswax.
- Paper plates.
- Doilies.
- Paper covering for the tables, or table cloths.
- Tiered staging saves a lot of horizontal space and is a far better method of displaying extracted honey. If it is painted white, it shows off the honey extremely well.
- Telephone numbers of venue key holders if relevant.
- Telephone numbers of the judges

4. Honey grading glasses (John Goodwin)

you have booked. Ensure that they have your contact numbers in case of journey difficulties.
- All returned trophies cleaned and polished.
- Non-returnable trophies (silver spoon and honey jar).
- Prize cards.
- Prize money.
- Lots and lots of pens, paper, stick-it notes and other stationery.
- Marking sheets and/or simple duplicate books.
- Furniture not supplied by the venue.
- Spare copies of the lists of items exhibited and the exhibitors for the show.

Method of allocating entries

The normal method of allocating entry numbers is by the use of a matrix table as shown.

	Class number						
Exhibitor Name	Light honey 1	Med honey 2	Dark Honey 3	Wax block 4	Mead dry 5	Mead sweet 6	Colour photograph 7
Mrs Smith		1		2			
Miss Jones			3	4			
Mr Goud	5	6/a					

The entry jar /exhibit slips for the three exhibitors would read:

Mrs Smith

| Class 2 | Entry number 1 |
| Class 4 | Entry number 2 |

Miss Jones

| Class 3 | Entry number 3 |
| Class 4 | Entry number 4 |

Mr Goud

Class 1	Entry number 5
Class 2	Entry number 6
Class 2	Entry number 6a

As will be noted, Mr Goud has a double entry for Class 6 so he will be supplied with two slips, one for entry 6 and one for entry 6a.

This method of numbering double entries makes it easier for the judge to recognise instantly that there is a double entry. The BBKA rules that the exhibitor can only win one prize per class. The judge will judge the two entries and pick the better of the two so that only one entry will go forward to be judged against all the other entries.

More experienced show managers might find the above explanation tiresome but hopefully it will be helpful to the less experienced.

This table can of course be automated by using a database program, or you can use a simple spreadsheet. If it is decided to use a computer program, it is very important that it is ultra-reliable, with spare hardware available in case of failure, and a backup procedure considered. A good plan is to carry out a number of dry runs. There should if possible be a large display screen that can be seen by all the exhibitors, so that the whole process is open and cannot be criticised for not having been an openly and fairly judged show.

Some of the more sophisticated software programs seen at some honey shows will automatically print the prize cards from the input results and run a tally for prize money for each successful exhibitor.

Prize cards should be of the best quality, whether computer-generated or handwritten. If handwritten, try and find a skilled calligrapher to add to the prize card's prestige. Exhibitors really only exhibit for the prize card, so they should be the best the association can afford. The prize card is the only award that the exhibitor walks away with, and their only real award for exhibiting and all their hard work preparing their exhibits. Long gone are the days when prize money at honey and other produce shows was the equivalent of an agricultural labourer's wages.

Layout of the classes

Never lay out a honey show as if it were a village produce show. Honey and its associated classes should always be shown on tiered racking, mead on racks, and photographs should be shown on stands at eye-level height. If there is a class for microscope slides, arrange them so that the public do not need to use a microscope. A good method is to photograph the slide via the microscope and display it in a slideshow on a large screen.

Hives: both nucleus and observation hives
    Observation hives are either exhibited with free-flying bees or the bees are enclosed.
    If the bees are enclosed, the observation hive should only be shown for the day.
    In the case of observation hives with free-flying bees, they must be against an outside wall with a pipe to allow the bees to fly outside as stated above. Hives should never ever be left without a steward, for the bees' safe keeping and public safety.
    Nucleus hives are shown outside, usually within a cage so that the worker bees have to fly above head height.

Honey should always be shown on tiered staging, so that no two entries are in front of each other. These tiers can be up to six exhibits high. The other advantage of showing honey on tiers or steps is that the upward space has no cost. A honey show hall is required to be long and thin, but needs to be very well lit. If exhibiting is on the same level on a bench, this takes up a tremendous amount of space, which can result in narrow corridors between the room's showbenches. This is not very satisfactory for a number of reasons: exhibits are very difficult to judge, it's very poor for the public as the honey cannot be seen very easily, and it's difficult for the steward to sort out the exhibits. Safety is also often an issue as big wide benches result in narrow corridors which are not good for wheelchair access and that of the general public.

Mead needs stands that enable the light to be shone through the bottles.

Photographs need to be shown on a stand, not flat on a table. Again, this needs a lot of wall space. Aisles and walkways need to to be wide enough to cater for the expected visitor numbers and wheelchairs. Obviously there are lot more visitors to a county agricultural show than a county association show.

Procedures for the show day

Firstly, make the show day a happy one. There is nothing worse than everyone being worried and upset and feeling under pressure. So, keep the atmosphere light with banter, but not forgetting that we have a serious task to implement over the next few hours.

Allow for a **stewards' rota** – what is often forgotten is that there is a requirement for manpower once the judging of exhibits is over. Stand stewards are still required. If the honey show is just a section of a larger show, it will be necessary to man the show possibly over a number of days: this can take a lot of steward manpower. One of the best ways to set up a rota is to allow the stewards to organise it themselves to suit their convenience. Nevertheless, the stand or section needs manning at all times of the show. If the show is part of a bigger show which runs over a few days, clarification of overnight security needs to be established between the honey show section and the management of the show.

Security of exhibits and safety. Where observation hives are on show, it is sometimes advisable to allow just a few of the public to some of the observation hives, due to safety and space requirements.

Also make sure that there is enough manpower for the putting up and breakdown of the show.

On the day of the show the show secretary should introduce the stewards and the judges, explaining their duties for the day. The judges' stewards may well be gaining experience as they work towards their honey judge exam. The judge should be encouraged to explain the judging procedures to their steward. If the steward has an entry in the show about to be judged, then that steward must inform the judge prior to the start of the judging and at no time make any comments on the exhibits.

If the show requires open judging (judging where the public have access and can observe the classes being judged), extra time and possibly judging resources need to be allowed as the judge and judge's steward will more often than not be expected to interact, discuss, explain to the public how honey and honey products are judged, and answer questions on general beekeeping.

Before judging commences, it is best to gather together all the personnel involved with the organisation and judging of the honey show for a briefing covering safety and fire procedures and the duties of the stewards. How do you want the judges to operate in regard to novices and open classes? At the briefing, it is worthwhile stating which judging is required to be carried out with marking sheets, and what testing equipment should be used – grading glasses, ISO wine glasses, etc. There are some shows now that do not allow knives (used for judging the confectionery classes) to be brought onto the showground, in which case the show manager should make sure that these knives are provided, and have a procedure for booking them in and out. When the candles are lit and are on a long burn time test, make sure that there is a steward in attendance at the burn at all times with the necessary equipment to extinguish a fire if there is an accident.

The show manager should issue marking sheets for classes that require them.

It is also a requirement that these marking sheets are returned, filled in, with the results sheets to the show teller or show manager.

It is the show manager's responsibility to allocate which classes he/she wants the judge to adjudicate, ensuring that the judge is qualified for the relevant classes. Most judges like to judge the exhibits in the order that suits them and the logical geographical order, and more importantly, in view of strength of taste of the relevant classes. A good example would be dry mead before sweet, and light honey before dark or heather honey. When planning order of judging, geography and order need to be considered so that the judges are not in each other's way.

If the judge decides that he will not award any prizes for a particular class, the show manager should be summoned by the judge to the display stand, enabling the judge to explain the reasons for his decision. The final say is the judge's. However, the secretary must be fully aware of the reasons and should request the judge to fill in a "not to schedule" or disqualification form.

The show secretary should, if at all possible, refrain from actually showing at the show they have responsibility for. Apart from a possible perception of impartiality, there are enough duties for the secretary prior to and during the show without the worry of looking after his or her exhibits. Try and appoint as many stewards as possible who are not showing. However, it is also a training ground for exhibitors and potential judges to get experience and knowledge of what the judges are looking at and for.

Depending on the size of show, it is sometimes wise to appoint a show judge referee, who is appointed to handle any disputes between exhibitor and the judge. Of course, the referee cannot in any circumstances have any entries in the show.

The show secretary should see that a water supply and washing facilities are available for the judges and stewards.

Always make sure that your judges and stewards are well fed and watered. Try to arrange for the stewards and judges to dine together: there is nothing worse than for the judge to have to say goodbye to his steward because the latter is not entitled to a lunch or tea.

Finally – very old fashioned but it does matter a lot – letters or emails of thanks should be sent on completion of the show project. Sponsors should always be thanked with a brief resumé of the show, including number of exhibits, classes and prize-winners; send letters also to your judges, judges' stewards and all other helpers with a resumé of the show covering entry numbers and show winners.

If there are gift classes whose exhibits have been promised to organisations, make sure these exhibits are delivered to them, preferably by a committee member or officer of the association. A photograph of the handover is a good idea: this can be passed on to the local press with a short written report. This is very good PR for beekeeping. Make sure your sponsors are aware of the gifting classes as they will appreciate knowing about the show they have sponsored and its profile in the local community.

Virtual honey shows

During the writing of this book in 2020, we have had virtually no honey shows all year: even the National Honey Show was cancelled due to the COVID-19 pandemic. However not all was lost – a couple of county associations managed to run virtual honey shows. The National Honey Show did not have a show but still managed to offer online lectures and workshops.

The virtual honey shows were run very successfully – the county associations that held and ran these honey shows were lucky enough to have the resources to use modern technology. Exhibitors sent their entries via the web to the show secretary who then passed them on to the judges.

The schedules were discussed at some length with the judges prior to publication. Wax and candles were offered as a proposed class – these were dropped after further discussions with the judges. Getting some judges to get their heads round what could be judged via the net was quite interesting.

The following classes are a sample of classes that can be judged online or virtually:

- A macro or close-up slide
- A photograph with descriptions of the subject
- Black and white picture of bees or beekeeping
- Photomicrograph of pollens
- Photomicrograph of honey bee anatomy
- Photomicrograph of honey bee pests or diseases
- A short video of beekeeping

Poetry, Essays and Articles
- A short article on beekeeping
- A press release for the local or national press
- A fun short article on beekeeping
- An article about one of your beekeeping experiences
- A short article with general public interest
- Relevant health and safety advice leaflet
- A limerick about bees or beekeeping

Art and Crafts
- A honey label
- A cake decorated with a bee or beekeeping theme
- A beeswax model
- A painting or artwork
- Any craft item made by the exhibitor with a bee or beekeeping theme
- Design a logo for a Beekeepers Association
- Design a pin badge for a Beekeepers Association

Junior classes
- Painting (with various age groups) of a beekeeping subject
- Handicrafts (with various age groups) on a beekeeping subject
- Essay (with various age groups) on beekeeping experience

2
Schedules

Show schedules should be written with no ambiguity so that both judges and exhibitors understand what is required.

As a show manager, you should be aware of some pitfalls of badly-thought-out show schedules, especially if they are linked to show entrance ticket concessions. In some of the bigger floral or agricultural shows, ticket concessions are linked to classes entered. This can lead, and has led, to instances where there have been up to 150 entries in the honey show's Confectionery section of cakes, biscuits and fudge. This number of entries can and did swamp the entries for the wax and honey.

If you become a show manager of a honey show that is just a show section of a larger show whose ticket offers can impinge on the honey section, make sure that the show directors are made aware of the implications of such offers. It is up to the honey show manager to lobby and make known how some badly-thought-out ticket offers can affect the honey show.

The above anomaly can happen as a number of classes are open to non-bee-keepers, by virtue of allowing honey and wax that are not from the exhibitors' bees to be used in exhibits. Examples of such classes are Beeswax Candles and Honey Confectionery. This ruling is intended to encourage more exhibitors to enter classes.

There are various arguments for and against this type of ruling in the schedule. The argument for allows greater numbers of exhibits and exhibitors, and encourages beekeepers who probably would not become exhibitors.

The argument against is that the honey show could become a cake and confectionery show.

Extracted honey

Always specify that no labels or markings be on the exhibits other than the labels provided by the show officers.

Some examples for reference for the following extracted honey classes:
- Light-coloured honey
- Medium-coloured honey
- Dark-coloured honey
- Heather honey
- Soft set honey
- Granulated honey

Layout of classes will depend on number of jars per exhibit. It is normally two jars, although novice classes might only require one jar.

A general layout of classes would be:
- 2 jars light
- 2 jars medium
- 2 jars dark
- 2 jars soft set
- 2 jars naturally granulated (not stirred)
- 2 jars Ling Heather honey
- 2 jars Heather Blend
- 2 jars chunk

Type of jar and lid should be stated in the schedule. Normally it would say that extracted honey must be exhibited in plain, clear 454g (1lb) squat jars (glass or plastic) with a standard gold-lacquered metal, or gold plastic, screw or twist-off lids and matching except where otherwise stated. In some shows, white plastic lids are accepted. The position of the exhibition label should be stated: normally it is to allow between 10-15mm between the edge of the label and the bottom of the jar or bottle.

Soft set or granulated honey should read "naturally granulated" or "soft set" and should always be a separate class.

Commercial classes

There is a need with these classes to specify not only number of jars but also:
- Jar size and jar type
- Are they restricted to a specific type and colour of honey?
- Labelling as for sale per sales regulations

Black jar class

This class is really a novice class. The contents of the jar are blacked out with gaffer tape, similar electrical insulation tape or a black paper sleeve. Exhibits are only judged on the aroma and taste. However, if on opening the jar the judge decides that the honey is not of a clean enough sample, they will not proceed to taste, so the exhibit will not be judged.

Comb honey

Frames for rotary extraction

The schedule would normally state: "One shallow or deep frame and/or one frame of honey, any size, to be suitable for rotary extraction." In this case, one would expect the frame to be wired and the honey liquid, otherwise it would not extract.

A frame of Ling Heather honey should be in a separate class. The schedule could read: "One frame of Ling Heather honey suitable for extraction or cut comb. The heather frame should not be wired if used for cut comb."

Specify that the frame is to be shown in a frame showcase and specify where the provided labels are to be placed on the showcase. It is advisable to state that plastic foundation is not allowed. The reasoning for including this in the schedule is that it is not possible to check for granulation: the plastic foundation is not transparent/ translucent, so the frame cannot be accessed for granulation or different colours of honey.

Cut comb

With cut comb honey, always allow a reasonable weight tolerance: plus or minus 10% is good guide. The following rules may be of use to suit the show being managed:
- Normally the class scheduling will request two pieces of comb.
- Specify where the labels provided are to be placed on the cut comb container or containers.
- Specify to be shown in a sealed commercially available container.
- Specify type of container.
- Always allow a reasonable weight tolerance: plus or minus 10% is good guide.
- State labelling requirements, both placement and the labels to be used (only labels issued by the show committee).

Sections

Always try and have separate classes for round sections and square sections. Examples of the type of container for showing the section are either a commercial cardboard box or a section showcase. Specify the weight expected, with a good tolerance, +/-10%

5. Wax in sealed containers (Gerry Collins)

Wax

Most judges will supply weigh scales, and should have a test weight traceable to NPL (National Physics Laboratory). It is prudent however for the show manager to be able to provide scales with a standard test weight and also go to the trouble of checking the scales provided by the show judges.

Beeswax blocks

In terms of scheduling, a single piece of wax to be exhibited should be in a sealed container (photo 5): a simple plastic bag, handmade wood or glass showcase, or a plastic cake dome (normally provided by the show committee). The reasoning for showing wax sealed in a container is twofold: the wax aroma is contained so the aroma can be sampled by the judge, and also beeswax exhibits open to the environment (especially during the summer months) attract honey bees and other

insects. A beeswax block schedule specification should consider the following rules to suit the show being managed:
- All beeswax blocks exhibited must be produced from the exhibitor's own apiary
- Weight of block to be specified
- The beeswax block to be shown polished or unpolished
- Dimensions of the block, specifically the depth
- Beeswax used for the showing of blocks must not be bleached
- All blocks to be exhibited in bee-proof containers of a size that can be specified
- Option of the show management supplying sealed containers (cake domes) and paper plates
- Beeswax block to be shown either patterned or plain cast (should be specified in the schedule)
- Labelling: position for placing the issued show slip or label to be specified. (Never ask an exhibitor to stick a label to the beeswax block as the label glue spoils the block finish.)
- Always allow a reasonable weight tolerance: plus or minus 10% is good guide.

One of the benefits of using a plastic dome and base plate issued by the show management is that all containers are the same, so no one can be accused of knowing who each wax exhibit belongs to.

28-Gram blocks of beeswax

In terms of scheduling, small wax blocks to be exhibited should be in a sealed container: a simple plastic bag, handmade wood or glass showcase, or a plastic cake dome (normally provided by the show committee). The reasoning for showing wax sealed in a container is two-fold: the wax aroma is contained so the aroma can be sampled by the judge, and also beeswax exhibits open to the environment (especially during the summer months) attract honey bees and other insects.

A beeswax block schedule should consider the following rules to suit the show being managed:
- All wax beeswax blocks exhibited must be produced from the exhibitor's own apiary.
- Beeswax blocks to be shown polished or unpolished
- Weight of blocks to be specified
- Specify the number of blocks. Remember that most moulds sold for casting 28g blocks have the capacity for five wax castings to be made at each casting process. Specifying six blocks to be exhibited needlessly increases the difficulty for the exhibitor in entering this class.
- Dimensions of the block, specifically the depth
- Beeswax used for the showing of blocks must not be bleached.
- Option of the show management to supply sealed containers (cake domes) and paper plates.

- All blocks to be exhibited in bee-proof containers of a size that can be specified
- Labelling: position for placing the issued show slip or label to be specified. (Never ask an exhibitor to stick a label to the beeswax block as the label glue spoils the block finish.)
- Always allow a reasonable weight tolerance: plus or minus 10% is a good guide.

One of the benefits of using a plastic dome and base plate issued by the show management is that all containers are the same, so no one can be accused of knowing who each wax exhibit belongs to.

6. Beeswax candles (Enid Brown)

Beeswax candles

A Beeswax Candle (photo 6) schedule should consider the following rules to suit the show being managed:
- Wax from sources other than the exhibitor's bees is normally allowed.
- Method of production required or allowed (poured, moulded, dipped and rolled)
- If the show is a major honey show and there are enough entries in the candle class, split the class so that different methods of casting are recognised by the judge.

- Dimensions of candle: length and diameter
- State that one candle will be lit if the class requires this
- If candleholders are required, state that they must be fireproof.
- If the show management is going to provide spike-type candleholders, the exhibitor should be informed.
- Labelling: position for placing the issued show slip or label to be specified. (Never ask an exhibitor to stick a label to the beeswax candle as the label glue spoils the candle finish.)

7. Beeswax fruit and vegetables (Joyce Nisbet)

Beeswax flowers, fruit and models

A Beeswax Model (photo 7) schedule should consider the following rules to suit the show being managed:
- Size of display and type of container to be used
- Number of pieces on display
- Difference between fruit and vegetable
- Judging of container or not
- Wax colouring allowed or not
- Exclusion of candles or not
- Additional material as props and structure assistance, e.g. steel or copper wire

Beeswax wraps

This class is quite a recent addition to honey show schedules and has attracted a number of new exhibitors to the show circuit. A beeswax wrap schedule should consider the following rules to suit the show being managed:
- Size and number of wrap pieces to be supplied
- Wax source must be stated to be of beeswax only.
- If there is an application of the wax wrap to be carried out during judging, this should be stated – for example to wrap a piece of foodstuff.
- Method of display (display board, etc.)

8. Shop window display (Gerry Collins)

Displays

There are three types of display that are scheduled at honey shows:
- Victorian pyramid
- Shop window (photo 8)
- Counter display

All three types of display class schedules should consider the following rules to suit the show being managed:

Material allowed
- Wax (coloured)
- Jarred honey (jar type)
- Cut comb
- Honey frames for extraction
- Sections
- Mead
- Beeswax polish
- Other products of the hive (cosmetic etc.)
- Minimum amount of honey to be displayed
- Minimum amount of wax to be displayed
- Base size of the display
- Height of the display
- For the counter and shop window displays, it is customary to have the honey and wax as for sale; labelling to be according to present legislation.
- Marks are also given for design of labels.
- For the Victorian pyramid display, there is sometimes a theme given.

Confectionery

All scheduled confectionery classes should consider the following rules to suit the show being managed:

Honey cakes

Use a recipe that enables the honey to be tasted.

If the honey type is not stated, a canny exhibitor will always use a darker or heather honey as the darker /heather honeys are more aromatic.

9. Cake display (Gerry Collins)

- What size of cake tin the exhibit should be baked in
- If not provided, specify the type and size of plate on which the cake or cakes have to be displayed.
- Try and provide plastic cake domes: they both look well and the judging and stewarding is a lot easier if plastic bags do not have to be opened and closed. Also, if a large cake is being shown, it can be shown by lifting one half of the cake over the other half (see photo 9) – this shows the dispersion of the fruit, and also an indication of the bake.

- Provide a cake base or paper plate.
- State method of labelling.
- State that the cake will be sampled and must be baked/produced to a standard suitable for sale to the public

10. Honey biscuits (Gerry Collins)

Honey biscuits

- Use a recipe that enables the honey to be tasted.
- If the honey type is not stated, a canny exhibitor will always use a darker or heather honey as the darker and heather honeys are more aromatic.
- State the sizes of biscuits to be shown.
- How many biscuits should be shown, or alternatively if all the biscuits produced from the scheduled recipe should be exhibited.
- If not provided, specify type and size of plate on which the biscuits have to be displayed.
- Try and provide plastic cake domes: they both look well and the judging and stewarding is a lot easier if plastic bags do not have to be opened and closed.
- State method of labelling.
- State that the biscuits will be sampled and must be produced to a standard suitable for sale to the public.

11. Display and list of flowers (Gerry Collins)

Bee forage (flowers)

This class is usually scheduled and exhibited at flower shows: one well-known flower show has a class called Bee Pasture. This is a very good class to add to a show, and very easy to schedule.

Typical wording would be:
- Twelve flowers displayed: all flowers are to be foraged by the honey bee.
- Check with show management on their policy on the showing of invasive species (e.g. Himalayan Balsam). Invasive species are banned from being exhibited at some shows.
- Vase size can be stipulated.
- Total display size
- List of flowers including Latin names and genus (if desired). (Photo 11).
- Flowers exhibited to be named and listed on an A4 sheet of paper encapsulated in a plastic covering
- The judges should have in their kit the listing published by the BBKA of all the forage plants. Prudence would suggest that the show manager has a copy.

Pot plant

A second class to have under Bee Forage would be a Pot Plant class which is foraged by honey bees; size stipulation can be written in the schedule. This is a very good class to add to a show and very easy to schedule.

Typical wording would be:
- Vase size
- Check with show management on their policy on the showing of invasive species (e.g. Himalayan Balsam) as such plants are banned from some floral shows.
- Must be a plant that the honey bee would forage
- Total plant size

Children's class

Can be an extension of the novice classes. There are a number of pitfalls with children's classes. Firstly, the class has to be designed so that a child has done the work to produce the exhibit. Secondly, it is advisable to have very narrow age bands for the class to be fair and judgeable. It may mean having a greater number of classes with few entries in each class. The schedule should always ask for the age of the exhibitor to be stated on the exhibit.

There is nothing better for promoting beekeeping than enabling a child to win a prize card, to encourage further exhibiting and taking up beekeeping in the future.

Cosmetics

If you intend to have classes with cosmetics, it would be prudent to make sure that you are aware of the relevant legislation. To market cosmetics in the UK at the present time, it is required that products for sale be compliant with the CPSR (Cosmetic Product Safety Report) and registered. So, if the cosmetic is to be labelled for sale, the exhibited product must have gone through the relevant procedure and assessments in order to be marketed and shown.

Invention

Inventions should be a practical device enabling a beekeeping task to be made easier:
- Size
- Any medium, including metalwork, woodwork, graphic design, documentation (to name just a few).
- Limit: not shown before
- Allow for description props

Products of the hive

In writing a schedule for products of the hive, some care must be taken. The show manager needs to have a picture of what is required to be shown. This is another very good class to add to a show and very easy to schedule.
A typical wording would be:
- Number of products to be shown
- Allowing (or not) the common products of the hive: honey, mead, and wax
- Size of display
- Minimum size of components
- If food products are to be included, the exhibitor must be informed that they will be tasted, so should be produced to a food hygiene standard.

Honey beer

The following schedule ideas may be useful:
- Normally shown in pairs or trios of bottles
- Bottle type: normal brown beer bottle with crown tops
- Bottles to have no imprinted ornamentation or commercial lettering on the glass
- Crown tops
- Filled to 15-20mm from the top
- Specify where the label supplied by the show should be located on the bottle
- Also state that the beer will be sampled

Honey vinegar

In most shows there is usually only a small entry for this class. The following note may be useful for your schedule:
- State amount of liquid vinegar that is to be provided
- Normally most shows will decant from the exhibitor's vessel into their own container. These containers are normal industrial laboratory-standard glass-stoppered bottles.
- The above task is normally carried out by the show stewards

Mead

Smaller shows normally have just sweet and dry mead. It is a good idea to supply racking which is backlit (photo 12), to show off the mead to its best advantage for the public. It is important to specify what is important for the mead classes.

12. Backlit racking (Enid Brown)

The following guidelines for the proposed schedule for your show may be useful:
- Bordeaux-style wine bottles of 70cl capacity, clear and colourless glass (slightly green-tinted from the glass moulding process to be allowed)
- Punted bottles with no imprinted ornamentation or commercial lettering on the glass
- White plastic flanged corks
- Filled to 15mm from the top
- No flavouring to be added
- Nothing to be added to the mead other than acids, nutrients and tannins
- Specify where the label supplied by the show should be located onto the exhibited bottle.
- Also state that the mead will be sampled.
- For all fruit-infused mead, notification to be made as to whether the exhibit is sweet or dry.

Microscope slides

This is a relatively new class, which has encouraged beekeepers with different skills from traditional exhibiting beekeepers to exhibit at show – which they perhaps would not have done in the past. If there is enough interest in your association for microscope classes, the class breakdown could be: diseases, pests, pollen and bee body histopathology.

The following guidelines for the proposed schedule for your show may be useful:
- Number of slides to be shown (can be a pair)
- Slide specification
- Method of description of slide content on slide or separate paperwork
- Subject matter
- If the slides are going to be exhibited by projection or some other medium to the public after judging is completed, the exhibitors should be informed in the schedule.

Photographs

If electronic submission is allowed, the show organisation needs to budget either for printing or for a good method of display that the public can utilise. Always include a disclaimer for copyright, as there is no easy means for the show committee to check the originality of the exhibit.

Below are some of the scheduling rules that you may want to apply:
- Subjects: honey bees, other species of bee, forage, products, and beekeeping.
- Could include photomicroscopy with varying subjects available with this technique.
- Size and format.
- Colour (photo 13) or black and white (photo 14).
- All the taking and producing of the photograph to be by the exhibitor. If using a film format, processing can be by a third party.
- Electronic submission.
- Time for submission at a timeslot some time prior to judging the conventional classes, if using electronic submission.
- Copyright: does the show want the right to use for promotional requirements for the benefit of the show at a later date?

13. Colour photograph taken at R.H.S. Flower Show (John Goodwin)

14. Black and white example (John Goodwin)

Videos

Always include a disclaimer for copyright as there is no easy means for the show committee to check the originality of the exhibit.

Below are some of the scheduling rules that you may want to apply:
- Subject
- All the taking, producing and directing of the video to be by the exhibitor
- Length of film /video run time
- Method or media stated to suit judges' and shows' resources
- Copyright: does the show want the right to use for promotional purposes for the benefit of the show at a later date?
- Normally, electronic submission
- If background music is used, certificate of authorisation should be submitted.
- Time for submission: normally a timeframe some time prior to judging the conventional classes

Honey label

If there is a class that causes controversy, this is the one and it takes a lot of time to judge at most shows. It is advisable for the show management to have a copy of the relevant legislation available for the judge.

Below are some of the scheduling rules that you may want to apply:
- Size of label to be shown on what weight of jar
- Shown on empty jar or on board either supplied by the show or the exhibitor
- What medium do you require? Is a download necessary?

There is a problem with just stating that the design is according to current labelling legislation, because labelling can be, and is, interpreted slightly by different trading standards officers working for different authorities. It may be prudent to lay down your own interpretation of the trading standard legislation for the show schedule.

An example may be useful:
- Name and residential address
- Produce of the country of origin
- Best before date, and lot number (normally suggest a timescale of 2-3 years)
- Metric weight in bold with font height of 4mm
- Imperial weight, not in bold, in proximity to the metric weight
- The word "honey" may be prefixed with a type (e.g. heather) or location

Essays

Always include a disclaimer for copyright as there is no easy means for the show committee to check the originality of the exhibit.

Below are some of the scheduling rules that you may want to apply:
- Subject and format could be a branch newsletter, or note to a fellow beekeeper, or essay – give the topic.
- Number of words
- Number of copies required
- Print or written
- Submission by electronic format
- If paper is used, suggest paper size specification
- Copyright: does the show want the right to use for promotional requirements for the benefit of the show at a later date? If so, this should be stated in the schedule.
- Time for submission for judging is normally a timeframe some time prior to judging the conventional classes.
- In larger shows, the essays are often sent to the judge prior to the show date. This allows the judge time to read the exhibit, make comments, and award a placing. The exhibits are thereafter displayed on show benches.

Handicrafts

As the term "handicrafts" can cover such a broad array of exhibits, it is worth going to the trouble of writing the schedule to cover the items that you require to be exhibited: for example, needlework, pottery and pictorial art.

Below are some of the scheduling rules that you may want to apply:
- Sometimes a theme is suggested.
- Size of exhibited work pieces
- Access to all sides of the exhibited work piece for judging purposes
- Separately-provided detailed description
- Handicrafts are normally broken down into "with or without needlework". Scheduling should include:
  - Medium: art, needlework (can specify what type of needlework: cross-stitch, tapestry), sculpture
  - Limit: to not shown before
  - Allow for description props.

Gift classes

The following guidelines for the proposed schedule for exhibited items at your show may be useful:
- State the charitable organisation that will benefit from the exhibited items in the gift classes.
- If a buyback of the exhibited items is allowed, a form that can be signed by the exhibitor should be part of the show rules, stating that the buy-back provision is be exercised.
- When the entries are being made with condition of buyback, the sum of money should be charged prior to the staging of the show exhibits.
- If the gift is going to be sold by the charity that benefits from the class, the exhibited items will have to be labelled as a For Sale class, so commercial legal labelling is required. To save having the show stewards re-labelling and fitting new tamper-proof labels, a good procedure is to ask the exhibitor to provide labels and tamperproof strips but not to put them on the jars. The fixing of labels and tamperproof strips can be fixed by the show stewards after the show breakdown prior to being handed over to the sale counter.

Observation hives

Observation hives are not seen at many honey shows due to most Beekeeping Association honey shows being held during the autumn (not a time of year suitable for moving bees), such as the National Honey Show when it would not be possible to set up an observation hive. The main showing of observation hives is not in a competitive environment, but as a public relations exercise at agricultural shows and flower shows. Unfortunately, a lot of these shows do not allow or cater for free-flying bees. Observation hives shown really have to be with the bees enclosed. The demonstration of enclosed bees means that the observation hive should be changed every day of the show. Also, although the public do not realise it, the bees do not behave completely naturally in a closed environment. Ethically it is questionable whether this type of demonstration should be allowed nowadays. However, there are still some shows where you get the chance to see a good showing of observation hives with free-flying bees. Always make sure that where the bees are located there is no chance of their overheating, that they are out of direct sunlight, and that the observation windows can be shut at night. It is worth writing into the schedule the organisation and arrangements for staging the observation hives and their break-down. A minimum arrangement would be for vehicle passes onto the showground enabling easy transport to the honey section of the show.

The following guidelines for the proposed schedule for exhibited hives at your show may be useful:
- It makes very good environmental, biological, security and safety sense to limit the bee classes to associations local to the show venue, if bees are free-flying.
- State if the bees will be in a situation where they can fly. Note: all bees should be allowed to fly if the show is to be held over more than twenty-four hours.
- If it is intended that the bees are to fly, the schedule should specify the pipe fittings required for the bee outlet.
- Physical dimensions of the observation hive: size limitation and number of frames.
- Specify that the bees are to be supplied with nutrition (syrup) and water.
- Specify marking and labelling for educational purposes.
- What is expected of the colony profile? (Brood, store and queen marked)
- Educational labelling: specify what is required over and above labelling. If required, an instruction sheet to be provided.
- Specify the method of strapping the hives and what straps the exhibitor has to provide.
- State method of installation of the observation hive and breakdown procedure after the show.
- There are a number of different types of observation hives, from the nucleus type where you just see one frame of bees, to the more traditional observation hive constructed as if it was a section of a double-brood colony or a section taken from the hive. With this hive you get two brood frames stacked vertically, a queen excluder, and then one or two super frames stacked again vertically on top of the queen excluder.
- Vehicle access to the show for the exhibitor has to be considered.
- Inform the exhibitor of the procedure in case of accidents.
- Lay down procedures for removal of the observation hive if disease is found.

Nucleus hives

Nucleus hives are usually shown in a cage positioned so that the public can view the bees safely. This should be a framed enclosure with high mesh sides and open top to allow the bees to fly. The high side encourages the bees to fly up and away from the public.

The following guidelines for the proposed schedule for exhibited hives at your show may be useful:
- Limit the entries to local associations' membership and/or the geographical area local to the show.
- Confirm the size of the frames, and how many, that are to be shown and confirm the model of hives allowed.
- What is expected of the colony profile? (Brood, store, and queen marked)
- Inform the prospective exhibitor that their bees may be used for demonstrations to the public, after judging has taken place.

- Specify marking and labelling of the queen for educational purposes.
- State the method of installation of the nucleus hive and its breakdown procedure after the show.
- Vehicle access to the show for the exhibitor has to be considered.
- Specify the method of strapping the hives and what straps the exhibitor has to provide.
- Inform the exhibitor of the procedure in case of accidents.
- Lay down procedures for removal of the nucleus hive if disease is found.
- What procedures are in place for the welfare of the bees? What is the exhibitor expected to supply? (Water bottles/syrup)
- Physical dimensions of the observation
- Number of frames
- Will all models of hives be acceptable (Smith, Langstroth, Dadant, variations on the national size, etc.)?

Skeps

Skeps are not a very common class these days. In the past it was one of the basic skills an amateur beekeeper would possess, which led to a large entry of skeps at honey shows.

The following guidelines for the proposed schedule for exhibited skeps at your show may be useful:
- Size of skep required
- Use of skep - catching swarms or keeping bees?
- Must be made by the exhibitor
- The exhibit must not have been shown before.
- Base plate or not
- Binding and material that the shown skep is to be made from
- Wattle cover to be shown separately (only seen in certain localities)
- Fitted with an eke
- Must be a new skep, not having been used for keeping bees, and free from brace comb.

3
Stewarding

There are a number of helping hands required at a honey show, a lot of them coming under the title of Show Steward.

If you get the opportunity, try and steward for a honey judge, even if it is at a small show. Stewarding is one of the best ways to learn a lot about showing.

Judge's steward

One of the best tips a prospective steward can take on board is to listen very carefully to the pre-show briefing so as to understand the custom, methods and procedures for the show, and what is expected of you. Procedures do vary such a lot between associations and organisations.

Make sure that you have a white laboratory coat, and judge's steward's trilby or similar outfit, disposable gloves (normally supplied by the show association), pen and notebook. If you do not have any of the above kit, please inform the show secretary that you require certain items. Again, this request should be made days, even weeks, before the show. The show secretary will have a lot of different tasks to perform on the show day so you will not be popular finding other tasks for a very busy person (the show secretary).

If you are on the path to becoming a judge and are stewarding the show to add to your portfolio, make the judge aware of your on-going interest in the judging procedures. When the judging of all the classes allocated to the judge is completed, ask if you can raise questions and have a discussion about judging methods and processes. Have the necessary documents with you ready for the judge to sign off. Very often the judge will have a number of documents to sign at the end of judging the show.

At some shows you may be acting as judge's steward while the judge is carrying out an assessment or exam for a prospective judge. Ask the judge what they will both require of you. You may be expected to act as judge's steward for the student judge. It is imperative that in this situation you keep your own council and remember that what you have witnessed is confidential and should not be discussed with anybody after the show.

Inform the judge if you have entries in the show and keep your own council when your items are being judged.

Make sure that the judge adjudicates on all the exhibits in the class that are to be judged. It is very easy for exhibits to get missed: usually this will happen at shows where one class runs into another on the tier racking due to lack of space or poor planning. A good stewarding method is to inform the judge how many entries there are in the class to be judged. The steward should have class sheets issued to them with all the entries for judging listed. If there are entries missing, and the sheet has not been updated, or you as the steward have not been informed, this should be raised with the show secretary or a similar show official before judging commences.

When the judge has completed the judging of a class and has decided the placings or winners, fill in the documentation with which the show officials have provided you. This documentation can be as simple as a duplicate or triplicate book, and/or a documented show form. It is imperative you ensure that the judge has signed off the judged class (at some shows it is expected that the judge sign against each placing, and you yourself as steward sign), and that you then return the documentation, by a runner or docket steward or return the form yourself back to the Secretary's desk. The class result returns should be done as soon as possible and not left to the end of the show.

There are a number of reasons why the steward has to be punctual in returning results to the show secretary's desk. These reasons can include an item with no recognisable number given a prize, the results being needed for allocation for another trophy or a mistake over prizes given to a double entry. These are just a few.

To sum up, the requirement of a judge's steward is to see all, hear all, and say nowt. Joking apart, if you have questions on why the judge does certain judging tasks a certain way, feel free to ask, especially if you intend to become a honey judge yourself.

Kit required
- Always wear while stewarding for a judge: a white laboratory coat, and hat (trilby or similar is used in the food industry).
- Disposable gloves (normally supplied by the show association)
- Pen and note book

4
The secrets of showing

Extracted honey

A suggested toolkit for showing extracted honey, over and above the kit that you use for normal honey production, which you may want to either purchase or borrow:
- A warming cabinet, the most useful bit of kit there is for a beekeeper exhibiting at honey shows. There are many uses for a warming cabinet when showing honey as well as beeswax, as you will see later on. A warming cabinet is also useful for the commercial production of honey and wax.
- A do-it-yourself cabinet is just as good as a purchased one. Construction is very easy: some of the best warming cabinets are made out of old butcher's or household fridges, modified and fitted with a 150 watt tungsten light bulb with an off-the-shelf thermostat (photo 15). Another possibility is to use a ceramic heater, as you cannot now purchase tungsten light bulbs. If you are using or building a warming cabinet, it is prudent to have a safety thermostat to cut out the electrical supply if the control thermostat locks on, causing the spoiling of the honey – and, in the worst-case scenario, a fire incident through overheating (photo 16).

15. Fridge with light bulb and thermostat (Frank Kirkham)

16 (top left). Safety thermostat.
17 (top right). Grading glasses.
18 (bottom left). Stainless steel filters.
19 (bottom right). Storage case.
All photos by John Goodwin.

Take care in using fans to circulate the air: you can have the problem of dust being circulated around the fridge, a particular problem when wax casting.
- Equally there are a number of commercially available warming cabinets that are also suitable for melting beeswax.
- Refractometer, perhaps borrowed from your association
- Magnifying glass, preferably illuminated
- B D grading glasses. Beg or borrow. All branches and associations should have their own grading glasses if they run a honey show (see photo 17).
- Small LED torch: there are a number of good, suitable LED torches.
- Honey creamer
- Ruler and depth jig for helping position show labels, made out of either tension wrapping or a piece of UPVC downspout cut to the depth to suit the show schedule instructions.
- Good quality 1lb jars with new lids – plastic or gold tin-plated (see notes further on).
- Stainless steel filter: the Thorne's double filter with sliders to fit on the top of honey buckets are really good value (see photo 18).

- Filter cloth 500microns, supplied by Thorne's. Other filter cloths may be available from other suppliers, but be careful they do not shed lint fibres.
- One of the best bits of kit ever seen for carrying jarred show honey is a storage case made for storing Christmas tree baubles, from Staples or other office suppliers. The cases have a series of plastic dividers that hold a single 1lb squat honey jar tightly and safely, keeping the jar perfectly upright without scratching or marking the jar. There are also polystyrene boxes that have compartments for six jars (see photo 19).

20. Consider the sizes of jars for exhibiting (John Goodwin)

Most classes are for a pair of jars of honey, so when exhibiting, make sure jars are from the same manufacturer – see photo 20 where all jars shown are different – and to the schedule's requirement. The normal requirement is for standard squat jars of 1lb capacity. This can vary as some shows now allow twist-lidded jars, and there are some classes that allow different capacity from the standard 1lb (commercial classes being a good example). The same mould numbers are not required but it looks better if they are. It also tells the judge that you are serious about showing. Try to find jars that are free from any blemish and make sure that there is no internal glass staining or green tinge which can give the impression of dirt in the honey. The best way to condition jars for showing is to wash them in a dishwasher using the most aggressive programme. This not only gives the jars a nice sheen but also shows up all the minor imperfections, enabling jars that are not optically perfect to be discarded. There are still jars in existence with a longer neck. Heather honey has a thixotropic consistency and is full of air bubbles. The longer neck allows for what would appear to be overfilling but heather honey, due to the air bubbles, requires this extra room. That way the exhibit will still weigh 454g.

Lids

Some shows now allow plastic lids: the National Honey Show does now allow gold-coloured plastic lids. If you are allowed by the schedule to use plastic lids, use

21. Varied colours of metal lids (Gerry Collins)

them, as they are easier for controlling the quality. Also, since a lot of exhibitors still use metal lids, when you come to collect your exhibits, yours will be easier to spot because not only will you have prize cards, you will have an entry where few people use plastic lids. A word of warning: if using plastic lids, make sure that they are from the same batch and manufacturer as they vary quite a lot in the depth of the lid. Make sure the lids are of the same depth and obviously the same colour.

If using metal lids, make sure they are the same tin-plating, as the colouring can vary considerably (see photo 21). It is better to use new lids for showing in the case of metal lids, as there is always the danger of the lid having been knocked or scratched, which leads to rusting. The plastic gasket in metal lids will often have a talcum powder film on it: this powder is used as the release medium in the manufacture of the rubber gasket. It has a very bad habit of dropping onto the honey surface, giving signs of dust in the honey, so always wipe the inside of the lid, checking at this stage for any signs of rust. Even new lids which have not been plated completely can show signs of rust in the threads. Always screw the lids on tightly at home, some time prior to travelling and staging.

Never attempt to replace lids at the showground, for two main reasons:
- Every time you remove the lid, you risk dust contamination on the surface of the honey. Summer agricultural shows are really bad for this as they can be very dusty if the weather is dry.
- Unscrewing the lid just before staging will inevitably cause a loss of aroma from the honey, and there will not be time for the aroma to fill the headspace prior to judging.

The following points are worth noting:
- Judges will not mark down for travel stain on the inside of lids. However, it is sensible to keep your jars as level as possible.
- Obviously, do not show dented lids.

The secrets of showing

Labelling

Make sure your label is fixed in the position that the schedule requires: this is normally a measurement from the bottom of the jar. A useful tool can be made to help guide the fixing of the label from a piece of plastic downspout cut to the length required for the label position or a piece of packing strap (see photos 22/23).

For a little bit of polish or showmanship, it is a good idea to place the label equally between the jar seams. Some shows allow booking in on the day: avoid this habit as the last thing you want on delivering your honey for staging is to be rushed. So always send your entries in early enough for the show secretary to return your showslips or labels back to you in time for you to fix all labels at home prior to delivery for staging.

22/23. A useful tool for labelling (John Goodwin)

Honey classification by colour

All honey shows have the following classifications for honey: light, medium and dark. These are classified with two standard B D Grading Glasses as shown (purchased from Thorne's). Light honey is lighter than the lighter of the two glasses, medium is darker than the light filter glass but lighter than the dark filter glass. It is prudent of the exhibitor to check his or her entries against the show manager's glass filters prior to entry. Make sure that the show manager has the recognised standard B D Grading Glasses (see photo 24).

The main honey crop produced in the UK is light honey. The main sources of this honey are brassica crops, OSR (oil seed rape), legumes (clover), and also sycamore. Sycamore is not a crop to be underestimated; it has been quoted that a sycamore tree produces as much nectar with the right conditions as four acres of OSR. (Some judges would suggest that sycamore

24. Using grading glasses (Gerry Collins)

honey could be classified as medium.) A later massive crop of nectar producing light honey can be from the lime tree, providing there is decent weather in July.

Medium honeys are obtained mainly from trees such as chestnut and hawthorn. Ivy also produces a very welcome late crop of medium honey. Ivy is not very often harvested as most beekeepers have fed their bees in the autumn by the time it flowers. Buckwheat, which is now grown as a game cover crop, produces very good crimson-coloured honey.

Under normal circumstance to get dark honey in the UK is quite an achievement. However, sometimes, through quirks of the weather and the position of the apiary, one can be very presently surprised and find that the bees have brought in dark honey. For the beekeeper who intends to show, this dark honey harvest is like gold.

Most dark honeys harvested are honeydew. The honeydew is harvested when least expected: it has been known to harvest honeydew when bees have been taken to the heather. What happened in one situation was that the bees found a forest station or plantation of soft woods which they worked, blending honeydew with heather honey and making a very aromatic and flavoursome honey. If you have produced honey in this situation, beware! Some schedules actually state that there should be no heather honey in the dark honey exhibit. How the judge could distinguish that the sample was a heather blend is not known.

Honeydew, which is the main source of dark honey, is secreted by aphids as they feed on the sap from various trees, both coniferous and deciduous, such as sycamore, spruce, chestnut, beech, cedar, ash, larch, willow and lime trees. This list is by no means the complete forage base. A complete list would be extremely extensive if all of the tree and plant and species worldwide were considered.

It is worth noting that honeydew is actually colourless: it is the pollution of sooty particles which makes for the dark colour.

It has been known for a novice beekeeper to be very concerned on finding a harvest of dark honey. So concerned was the novice he asked advice from seasoned exhibiting beekeeper/honey showman, asking what was the problem with his bees, thinking that the bees had got a disease. The seasoned beekeeper who was a renowned honey show exhibitor, told the novice that he should dump this dark honey as it was probably molasses that the bees had picked up from a cattle feed block. Obviously, the seasoned beekeeper was playing a very hard showman's game knowing that the novice beekeeper's dark honey was of show-winning quality. So, novices, beware in similar situations and get a second opinion. The honey went on to win numerous prizes at shows all over the country.

As dark honey is very rare and precious, the same exhibit will often will be shown for years. If it is not looked after, it will degrade over time. A good method of keeping your dark honey in show condition is to blend it with a lighter honey at the start of the show season. One method is to add a little medium honey to your dark honey and a little light honey (preferably this year's crop) to your medium honey.

One of the easiest methods for blending honey is to carefully warm your dark show honey and some medium show honey, then decant into a large container, mix together carefully, adding just a little medium honey to save re-blending, and re-jar into clean show honey jars. Store and re-skim ready for the show. This process is also used for freshening up medium honey, using light honey from this year's crop and giving the honey a fresh aroma.

Check with your grading glasses to make sure that the dark and medium honeys conform to the class in which you intend to show them. If they don't conform to the colour band, re-blend. Unfortunately this can be a bit of a long-winded process but you need to check the colour after each blend. The most important tip to be given in this section is to aim to blend to the middle of the desired colour range. If possible, never show honey on the border of a colour range.

Production

To show honey, start by looking at your hives: you need to keep your bees clean, on a clean foundation both in the brood box and in the super. It is also prudent to set up your supers with appropriate castellated spacers and suitable foundation. If you intend to produce comb honey, you will need to purchase thin unwired foundation. The other option for producing comb honey is to provide starter strips in the frames and allow the bees to draw out the comb. This takes a lot of management as you have to start with narrow spacing and move frames into supers with wider spacings. The changing of supers with various spacings has to be done in the production of comb honey, no matter which method you use. Allowing the bees to draw out their own comb is very time-consuming for the beekeeper and the bees.

When carrying out varroa treatment, be careful what treatment you use. Bees that have been treated with some remedies taint the wax and honey. (Of course, this taint could possibly come from another source of contamination.) Somehow the smell seems to taint both wax and honey, especially comb honey.

Extraction

For show honey, prior to actual extraction go through your supers. If you want to use any of these supers for showing, check that they are all capped. Even when the honey is fully capped, it may not necessarily have the right moisture content. If the bees have been working an extremely good flow and the weather has been hot and dry, the bees will cap honey that is still quite wet, with a moisture content above the legal requirement of 20% and 23% for heather honey. If you have any doubt about your honey moisture content being above the legal requirement, get it checked with a refractometer prior to storage as the honey will ferment. If your honey seems to be very runny, there is no point in showing it. The only honey that should go to a show is your very best with no known faults, because if you think there is a fault with your honey the judge will certainly find it.

When inspecting your frames, see if you have a good frame fully capped that would be suitable for showing as a frame for extraction. Many times, brilliant frames suitable for showing have been extracted. When it has been explained to beekeepers that they have got a possible show-winner prior to extraction these frames have gone to win many prizes. More than once beekeepers have said at shows that they have already extracted better frames than the one being inspected and discussed. So, the lesson is, if you think you have a good frame for showing, put it on one side and get a second opinion before thinking of extracting it.

A good method of uncapping is to use a long-bladed knife. Sometimes when extracting it is prudent to have two production lines. So, the honey for showing will be uncapped, being careful not to uncap any pollen cells as this leads to dull honey. If you have time, extract the show honey by dripping into a bucket, suspending the uncapped frames above a stainless steel bucket. You have to take care to do this in a dry atmosphere – like an old fridge – so that you can control the environment and keep the ambient environment warm but dust-free.

Another method used for showing honey is to spoon the honey off the midrib and then warm the honey to filter through a show honey cloth; this gives the best results for storing the honey cleaned and filtered.

Show preparation

Show honey should be crystal clear with no bubbles or scum. Even dark honey, when backlit, should show clear and be the colour of claret wine. A lot of honey is disqualified because of granulation: this again should be checked for with a torch by the exhibitor. Really clean honey takes a while to granulate. To achieve no bubbles in the honey, jarring up should take place some days or weeks prior to the show. Overfilling the jar almost to the top allows for the honey to be scraped clear of the bubbles. If the honey is warmed above 30 degrees centigrade, bubbles will normally float to the top to be skimmed off. If you have a reluctant bubble you could either use a cocktail stick to burst it, or a drinking straw – by sucking near to the bubble you can normally collapse it.

The jars should not be overfilled but when a lit torch is placed behind the lidded jar, there should not be an air gap. This is stated in the National Honey Show rules but not in the BBKA rules. At present there is no other easy method to ascertain that the correct net weight of honey is being exhibited, as different makes of jar can vary slightly in weight. Remember that the show schedule does not mention a weight of honey, only that it is shown in a particular container or jar.

Jars should be filled to very near the rim, thus allowing for honey to be skimmed until there are no bubbles or scum.

When jarring honey, it is best to use a commercial ripener (bottling drum with a high height to small diameter) which has the effect of pushing light contaminates to the top (e.g. bubbles etc.) by pressure. Usually take the middle of the batch for showing.

So, to recap:
- Keep the best-tasting honey on one side for showing: extract it separately from your commercial honey.
- Filter straight after extraction with a coarse filter.
- Warm slightly to filter with your fine filter cloth.
- Fill your ripener (a tall stainless steel honey bottling tank with a relatively small diameter) to get a head of pressure which enables bubbles to rise to the top.

- Leave overnight before overfilling your jars, checking that the jars are from the same manufacturer. (A good policy is to only use one brand of jar. If there is only one brand on your premises, there is no chance of mixing the jars.)
- Fill your blemish-free well-polished jars at an angle to alleviate the incursion of bubbles.
- Keep filled jars warm, skimming off bubbles over the next few days.
- Label up to schedule using the tool jigs made out of UPVC downspout or packing straps (see photos 22/23).
- Double-check your jarred honey with the grading glasses to make sure you enter the right class. Again, avoid showing honey near the colour boundary (see photo 24).
- Last, but by no means least, give your jars a last wipe to remove all finger marks.

Soft set honey

Kit required
- Warming cabinet as for the liquid honey.
- A mixer capable of continuous running for producing soft set honey. The mixing paddle needs to be run at quite a slow speed. You don't want the mixer to fold too much air into the honey.
- The most important bit of kit for the following two classes is a good quality magnifying glass, used for checking bit contamination of the exhibits (see note further on).

Soft set honey is manufactured by the Dyce process, which was invented and developed by Dr Elton James Dyce early in the twentieth century. The true Dyce process was developed to produce a spreadable granulated honey that did not ferment, so part of the original process included pasteurisation. This included the seed honey and the main batch honey.

As you know, honey with a high moisture content will ferment during or on completion of the granulation process. This is due to only a small amount of moisture being taken up by the crystals. The remaining part of the batch is then of a higher moisture content which provides a very good solution for fermentation to occur. One of the by-products of fermentation is CO_2 which can cause the honey to explode in the container, whether a jar or a storage vessel such as a bucket.

There are various methods used by amateur beekeepers to process their honey but the main points are: find a sample of very fine granulated honey – by fine, I mean that on tasting the sample of honey, there will not be a trace of grittiness to the taste or top of the mouth. Most importantly, this honey will have been filtered prior to granulation through a fine filter cloth (this step is very important – the last thing you want is on completion of jarring is to find bits in the jar). So, find a sample of very good aromatic and pleasant-tasting honey in a liquid form that has been filtered through a fine filter cloth.

The process

- Well before the following process is started, both the seed honey and the liquid honey should have been filtered through a fine cloth (the honey being used as the seeding prior to granulation).
- Melt the granulated honey until it has the look or consistency of porridge.
- Add to the liquid honey (which needs to be at room temperature).
- The ratio between the granulated honey and the liquid honey should be 10 to 1, that is, ten parts liquid to one part seed honey. This ratio is only a guideline – some people use virtually 50 per cent seed honey or even more up to 100 per cent. The less seeded honey you have, the longer it takes to disperse the crystals to give a homogenous batch.
- If you decide just to use the seeding honey, it is necessary to melt a portion to a runny clear state and then back-blend it with the remaining seeded honey into a porridge state.
- Use a mixer with an attachment suitable for creaming. The options are: hand stirrer, creamer blade to fit into an electric power drill, a fully automated creamer (see photos 25a and 25b), or a kitchen food mixer. (Be aware that the latter are not made for continuous running and too many people have ended up buying a new food mixer.)
- Keep creaming over a couple of days and keep skimming off the scum that rises to the top.
- Allow the scum to settle a little while prior to jarring up, to ensure it is all cleaned off.
- Jar up with plenty of time prior to the show, as when judged the honey should not move if the honey jar is placed on its side with the lid off.
- Keep checking your jars for bubbles rising which leads to scum.
- Then use a hot teaspoon that has been warmed in a cup of warm water and then dried. The heat of the spoon seems to remove the scum without picking up the honey

25a/25b. Automated mixer (John Goodwin)

- When all the scum has been removed, a quick spin with the hot spoon leaves a very nice show finish to the top of the honey.
- Soft set honey should have a matt dry surface, no movement and the consistency of spreadable butter.
- Finally turn the honey over and look at the bottom of the jar with a magnifying glass to check that there are no black bits in the bottom of the jar. Some people think that there is no harm in these black bits (thinking they are pollen) but cases have been analysed and shown that they were debris from insects. No further details are necessary (see photo 26 by Gerry Collins).

26. Checking the honey through the bottom of the jar (Gerry Collins)

- Beware of saving exhibits for the following year's show's soft-set honey class, as the honey can have a tendency to go back to a hard granulation. It does not happen with all honeys, just with some.
- Check your jars: make sure the same manufacture, the lids match, and not rusty. For further checks read the notes under Containers above.
- Label up show slips as per schedule.
- To be of show quality the honey should not run out of the jar when the jar is turned onto its side with the lid removed.

Granulated honey

This class is often referred to as Naturally Crystallised (not stirred).

Firstly, take some runny honey – preferably light honey. If you use some darker honey, the granulated colour is not that pleasing to the eye. This is a personal preference – there are no rules saying you must use light honey but that is the convention and most show people do use light honey. The honey should be of a show standard that you know granulates quite quickly. By "show standard" is meant honey that has been filtered and is very clean, jarred up in jars as discussed in the Extracted Honey section.

The honey needs to be completely granulated for showing, absolutely clean with no bits showing on the outside of the jars. Bits always seem to migrate to the bottom of the jars (see photo 27 by Gerry Collins): as stated under Soft Set Honey, some these black bits have come from a disgusting source.

Granulated honey will very often frost. In this, the honey moves from the jar side and the resulting air pocket shows up as a frost-like structure on the side of the jar. Frosting is not normally marked down. A good trick to stop frosting is to place the jars to be exhibited in the fridge for a couple of weeks just prior to completion of granulation, and remove from the fridge some days prior to the show. This seems to stop frosting in most cases. This process is due to the honey increasing in volume and expanding on warming, which puts it in contact under pressure with the jar side, stopping the honey from frosting. This tip works sometimes but not always.

Another tip to prevent frosting that sometimes works is to meld your granulated honey slightly in a large container until porridge-like, and then fill to the top the honey jars you intend to show with the porridge-like honey. Fasten down the lids. Keep turning the honey in a cool dry area. Normally this honey will not frost. Please note you need to allow a lot of time before the show for this process to work satisfactorily. The honey top will probably need skimming prior to the show.

A good tip is to make sure that your granulated honey has a feeling of granulation but not grittiness when tasted. If the honey has too fine a crystal structure, it can be thought to be soft set that is going back. The texture will end up like cold lard.

Re-cap:
- Check your jars – make sure the same manufacture; the lids match, and not rusty. For further checks read the notes under Containers above.
- Label up show slips as per schedule.

Black jar class

If there is an open class for the black jar class, it is worth exhibiting if you have a spare jar of honey. This is normally for runny honey judged just on aroma and taste. So pick a very aromatic honey with a good taste: the hygiene of your exhibit should be as for all your other runny honey exhibits.

Commercial class

This is one of the premier classes (see photo 27). It is usually a multi-jar class (e.g. 12, 10 or 6 jars) and can be for varying jar capacity: 42g, 113g, 227g, 340g, 454g and a mix of any of these jar sizes.

There are two unique points to this class:
- As it is a commercial class, varying types of jars are allowed, but always check the rules in case there are restrictions on jar size.
- Own labels compliant to current trading standards regulations are used as if the honey was for sale.

27. Commercial class exhibits (Gerry Collins)

When showing in the commercial class, you are normally required to use a label as for sale, which will have to conform to the relevant trading standards. The labelling and composition of honey for sale to consumers and catering establishments is controlled by (at the time of writing) by The Honey (England) Regulations 2003, The Food Labelling Regulations 1996, The Weights and Measures Act 1985 and The Packaged Goods Regulations 2006.

The description of the honey must follow one of the following reserved descriptions: Honey, Chunk Honey, Blossom Honey, Heather Honey, Cut Comb Honey, Honeydew. The term Raw Honey should not be used.

The following must be shown: Lot/Best Before date, country of origin, address of producer and the weight of the contents in bold. If showing imperial weight, the equivalent metric weight must be more prominent, or if the same type is used, the metric weight must come first, usually in bold vs a normal font for the imperial weight. The font size for net weight of one pound of product should be no less than 4mm high.

There must be a Best Before date. If day, month, year, then a Lot number is not essential. If the Best Before date is shown as end of month, year or end of year, then there must be a Lot number. This is all about traceability. All should be within eye view when the jar is on the shelf. "See base of jar" can also be shown on the label – this indicates where the BB date and Lot number are.

The label can also say "Surrey Honey" or whatever region it's from, but then must also have "Produce of UK".

Please bear in mind that these regulations are as when this document was typed. Regulations do change, so before entering this class not only read the schedule well, but take the trouble to read the relevant and up-to-date regulations.

When purchasing your labels, don't take for granted that the labels are according to the regulations. There was a well-known case some years ago of an exhibitor at the National Honey Show's commercial class being disqualified because the labels on their honey jars didn't have the required font size for the jar capacity being shown. These labels had been purchased from a well-known nationally trading beekeeping supplier.

Depending on the schedule, this class allows different jar sizes to be exhibited, which reflects the commercial sale of honey. The market for honey nowadays requires that a large percentage of honey sales is in different-sized containers from the standard 1lb squat jar with gold screw lids. Twist lids are also allowed to be exhibited on these different-sized jars. The use of twist lids leads to a major showing mistake by exhibitors, because the fill line for twist-lid jars is usually a lot lower than the bottom of the twist-on lid, resulting in a gap when the judge torches the exhibit. Again, read the schedule. The National Honey Show insist that the fill does not allow a gap between the top of the honey fill level and the bottom of the closed lid. So if the exhibitor just fills to the conventional fill line of a twist lid jar, they will be disqualified. It is prudent therefore to overfill the jar so that there is no gap. At the end of the day the exhibitor is there to win, so filling or overfilling takes away one reason for not winning.

The same care has to be taken when showing in the commercial class as other extracted honey classes so:

- Lids: follow the guide-lines above for the other extracted honey classes: no rust or dents, same plating, and use plastic if the schedule allows.
- In large classes, including commercial, all jars should be from the same manufacturer. It shows the judge that you really mean business and are exhibiting to win.
- Read the schedule carefully as the requirements for some commercial classes take a bit of understanding. Some classes allow for granulated and soft set, some for different jar sizes and stipulate how many jars and sizes of each are required.
- Tamper-proof strips: it is not a commercial requirement at the time of writing to have these on your jar. However, some shows do specify tamper-proof strips on the exhibited jars. Some of the smarter shows will request that a sample of tamper-proof strips is supplied but not fitted or glued onto the jar. It can be a little soul-destroying to have all your jars opened and the strips all torn without a prize. Looking on the bright side, if you haven't won a prize and all your jars have been opened, it does mean your exhibit was good enough to get to the final choice, but were put out for a minor fault, possibly on a single jar, with a fault on the outside of the jar.
- Always have a final check that the label content matches the jar content. This is quite a common fault found when judging the commercial class.
- If the show allows, always try to stage your own exhibits in this class. This is viewed as such a premier class that the winner very often also takes the Best in Show (Blue Ribbon).

Chunk honey

Chunk honey (see photo 28) is really a crossover product from comb honey and extracted liquid honey. The preparation of chunk honey needs to be carried out as short a time as possible prior to the honey show. The schedule needs to be read very carefully because there are a number of tips that if applied can give the exhibitor an edge if allowed.

If allowed in the schedule, use heather honey for the comb piece, if you have any available, as it will not granulate so easily. Heather comb honey will still granulate when used in the production of chunk honey, due to the interface with runny honey used as the liquid honey source, but not as quickly as comb non-heather floral honey.

Assembly of the parts to show chunk honey

Care must be taken with the lids and jars for the chunk honey class, just as for other extracted honey classes. Please read the description and help tips in the showing of liquid honey. Chunk honey is normally shown as a pair so the same care is needed as for other extracted honey exhibiting.

28. Chunk honey (Gerry Collins)

There are a number of methods of showing in this class. The convention is that the comb honey piece or pieces should be half the amount of honey – so half liquid honey, half comb.

Normally exhibitors will just use one piece of comb that goes right from the top of the jar to the bottom vertically (literally to just under the lid). It should be the width of the neck of the jar.

After cutting the comb, carefully place it on a cake cooling rack for draining, with the cake rack on a plate so that all the drained honey is saved. By allowing the comb to drain, the resultant piece will be relatively dry and any loose bits of cappings can be removed.

When dry, go over the comb very thoroughly to make sure there are no loose pieces of cappings or signs of pollen. All loose pieces must be removed very carefully. If the loose cappings are not removed, they will float about in the jar after assembly. Also check thoroughly for wildlife.

- To fix the comb to the bottom of the jar, one method used is to warm the bottom of the jar on a hotplate, taking not to overheat which could lead to cracking the jar or burning yourself. Remove the jar from the hotplate and place the vertical comb upright in the jar. The warm bottom of the jar will weld the comb to it. Check that the comb is perfectly vertical. The comb will now stay vertical and not float up or go diagonal in the jar when the liquid honey is added.
- The liquid honey component for the chunk honey class should be as clear and as bright as possible. A lighter honey is preferred to a darker one as the lighter honey will show off the comb honey better. The honey has to be as clean and as blemish-free as your extracted liquid honey with no bits or bubbly scum.
- Have the runny honey in a good pouring jug.
- The liquid honey is now poured into the jar very slowly, trying not to end up with any bubbles or disturbing the comb honey too much. If one piece of comb is being used, the fill is to the top of the jar to allow for skimming of any parts prior to the show.
- Top up if required with the liquid honey.
- Carriage: a great deal of care needs to be taken in transporting chunk honey. The jars need to be kept upright so there is no strain on the comb weld to the jar. One method to use a carrying-case as shown (see photo on page 50).

Showing

Try if possible to handle this honey yourself at staging. As with the runny honey, ensure there are no fingermarks on your jars and that the container part of the show exhibit is perfect – the last thing you want is your exhibit marked down or disqualified for a simple fault. Chunk honey takes an awful amount of time and preparation to show and is possibly one of the hardest classes to prepare. This is the reason that there are normally very low entries for chunk honey – which of course gives a chance of a prize card.

Storage

The problem with chunk honey is that it soon granulates, so the best advice is to sell or consume your chunk honey show exhibit as soon as the show is over. It will not be suitable for the next show, unless the next show is within a few days of the last one.

Gift class

For showing purposes, the Gift class should be treated and prepared for as the Commercial class discussed above. If you do not want to gift your produce, there is normally a buyback clause, which is often less than the value of the exhibit. The Gift class will be judged like the Commercial class so all labelling has to conform to the latest honey sales legislation. The schedule should specify that tamper-proof strips are provided but not stuck onto the jar: this enables the tamper strips to be fitted or stuck to the honey jars prior to gifting to the organisation or selling for association funds. If you can afford it and have the heart to gift your honey to a good cause, there is no better way of doing good and promoting beekeeping in the community. Make sure that the rules of buyback are very clear and what the funds from the buyback will be used for.

29. Composite class honey (Gerry Collins)

Composite class

If you have all the available components that the schedule requires for the Composite class, it is a good class to enter. There are usually a low number of exhibits entered for Composite classes, so there is always a good chance of a prize. The Composite class can be for a number of different items that are exhibited in a honey show: extracted different-coloured honey, a piece of comb honey, a wax block, candle and mead (see photo 29). Providing all the items are entered as scheduled, there is not normally any disqualification as this class is marked by summation rather than elimination and should always be marked with a marking sheet, which can be provided by the show committee, or the judge will have his own marking sheets. (For a copy of a typical composite marking sheet, see under the judging section of this book.) On completion of judging, the marking sheets are submitted to the show secretary and should be made available for the exhibitors to see at the end of the class judging, or when the show is open to the public.

To win this class or to get a place card, care has to be taken as above in Extracted Honey, Granulated Honey, Comb Honey, Wax or whatever the schedule has included. Each item should be checked over thoroughly as you would for the individual classes. Do not in any circumstance enter for this class any entries that have failed your quality control for the individual class. Although you cannot be disqualified for a particular entry, you cannot win with any item that negates the marks of the other items shown.

If the same size of jar is used, then all must be from the same manufacturer. If we are exhibiting to win, every stop should be pulled out to have a very professional display. In fact, the Composite class could be classed as mini-display, if the class has pieces to exhibit other than extracted honey. Containers should be as good as for the class these items would be shown in. For beeswax, make sure that the container or box is bee-proof, clean and easily opened. Comb honey should be in a clean display case that can be easily opened by the judge and is free from wildlife. Mead should be shown in the bottle-type requested, and make sure it is the variety specified by the schedule.

When packing for the show, a bit of advice is to pack all the components for this class together, checking that all components have been packed.

Heather honey

Heather is a crop found in more counties in the UK than you would think: from the heathlands on the south coast and Breckland in the eastern counties to the chases of the Midlands and the grouse and un-shot moors of Scotland, the West Country, North Wales, South Wales, the Peak District and the Yorkshire moors.

Heather honey is normally a separate classification in honey shows, and is usually shown in the following four forms: extracted, cut comb, sections and a frame for extraction.

Heather honey should never granulate. If your heather honey shows signs of granulation, it is possibly a blend. The common blend or contamination until a few years ago was rosebay willow herb which grows quite profusely on the in-bye land on the edge of the moorlands. The other floral sources which can cause contamination to the heather honey are white clover, ragwort, knapweed, thistle and bramble, to name but a few. Some of these plants do not actually grow on the heather land but on the in-bye land which runs up the moorland. The in-bye land has normally been heavily grazed by sheep, which helps encourage different floral species.

Most of these items of heather nectar contamination can be alleviated by taking your bees a little later in the month to the heather apiary. Remembering that most plants flower later at the higher altitudes where we find the heather moorlands.

One of the problems that have occurred on some of the prime sites that beekeepers have used for years is incursion of Himalayan Balsam along the streams on some of the moors.

Himalayan Balsam is the modern floral menace source at the heather apiary. You could leave taking your bees till well after the heather flow and your bees would still find Himalayan Balsam. This plant has become quite a pest, although welcomed by northern beekeepers for its flow relatively late in the season – it has been and is ruining a number of good moorland apiary sites. These sites produced top quality heather show honey year on year. The Himalayan Balsam plant seeds itself along riverbanks enabling the plant to colonise further upstream, colonising the river bank as it goes up to the higher altitudes, where we find the heather moor. Balsam has been found in the Peak District at over 1200 feet above sea level. The harsh weather and frost that are winter's norm for the uplands do not seem to affect or give a winter kill of Himalayan Balsam.

Another problem with the heather moors these days is a strong move to stop the rotational heather burning on grouse moors. Burning produces young heather which yields high levels of nectar compared with the long rank heather on un-burned moors.

The honey blend of Himalayan Balsam and Heather is very good-tasting but is no use for showing. We are of course discussing the sourcing of show honey – the above blends are not only perfectly good honey but very delicious. Contamination of these sources leads to granulation and small lumps with extracted heather honey, termed "bricks". The granulation and bricks are not acceptable for showing.

To prepare your hives for the heather you need:
- Very strong colonies with this year's queens.
- The colony should be manipulated so all sealed brood is in the centre of the brood box with the younger brood towards the outside. This means the sealed brood will emerge first, encouraging the queen to lay up the centre frames and that way all honey will go into the supers.
- This method ensures a mega strong colony in the brood box.
- Care has to be taken with this method: if the weather turns inclement the bees can very easily starve.
- Brother Adam at Buckfast primarily had his bees on Dartmoor and used Dadant hives for his heather honey production. His success was due to having large colonies and no shortage of fellow monks to help. Dadant hives are virtually twice the capacity of the national hive. They were fully up to strength just when the heather flow started, enabling him to achieve the high yields of heather honey.
- Do not over super on the heather. The bees are normally going back in population (brood shrinking) so unlike the setup of hives in the spring where we over super to help prevent swarms on the heather, you under super so that the brood can keep warm.
- The heather flow normally only lasts for a very short period of time.

Extracted heather honey

There are a number of ways to extract heather honey. In theory you can extract heather honey using a tangential rotary extractor after agitating the honey in the cells. Most beekeepers scrape off the cells into a bucket and then wrap the resultant honey wax mix in a linen cloth or similar textile material and press the heather honey using a heather press. Over-pressing the heather honey can sometimes give a waxy taste.

The more recent method is to scrape out as if you were pressing the heather honey but use a cappings centrifuge rather than a press. Although very expensive, the capping centrifuge does produce the best results by far.

Showing extracted heather honey: this is one honey where we have bubbles. Extracted heather honey is thixotropic. (Thixotropic material has two states. If energy (stirring) is applied to the jelly-like honey it will become less viscous and virtually liquid. This state is completely reversible when the energy is removed (stop stirring)). (See photo 30).

30. Extracted heather honey (Gerry Collins)

As with other extracted honeys, care should be taken with the container (jars and lids). Both should be absolutely spotless and free from glass stains, blemish, metal rust and dents.

To fill the jars of heather honey always overfill, keep in a dark cool environment and use a carry-box as you would for other jarred honey entrants. Slightly overfilling the jar with heather honey allows for the air bubbles – you will therefore achieve the correct weight.

Heather section

As with the production of sections from other floral sources, see notes on section production on page 73. The added problem with heather section production is that not only do you need a very strong colony, you have to be aware that if weather conditions change, which can happen very quickly on the heather moors, colonies can go from very strong to starving. To alleviate this, beware of taking your bees to the heather without some stores, especially if you have to travel a long way to your heather apiaries. It has been known for bees taken to the heather to be so strong and full of store that one person could hardly lift a colony from the vehicle to the apiary stand. On returning to the apiary two weeks later, the same bee colony could be lifted one-handed and two colonies carried back to the vehicle quite easily.

Do not be tempted to give your bees a lot of space when producing heather sections. If you are lucky enough to find on an apiary inspection that the bees have drawn, filled and capped a rack, do not be tempted to fit a second rack. The best plan is to clear the rack and fit a conventionally-framed super. The last thing you require, having gone to all the trouble of producing heather sections for showing, is to end up with travel stains and bees moving honey from the section by uncapping.

Heather frame for extraction

Normally the schedule would say "One Frame of Ling Heather".

A comb for extraction should always be shown in a bee-proof container with glass windows so that the exhibit can be seen by the show visitor. Timber for the container needs to be of a non-aromatic wood, such as hardwood: never pine, cedar or larch or any wood that gives aromas as this can taint the honey both for aroma and taste.

One point that should be made is that if you have ended up with a really good unwired frame of heather honey, that was to have been used for cut comb, it can of course be entered into the Frame for Extraction class. You do not need wired foundation for extraction of heather honey, as you do not extract heather honey with a rotary extractor. (This is not of course completely true as you could excite the honey cells and extract with a tangential extractor.) However, for show heather honey scraping and pressing is a far gentler process.

Please take care to read the schedule notes for a Frame for Extraction from other floral sources. Check for wildlife. If the frame is unwired, keep it away from any vibration and perhaps carefully store in the freezer until near the show date.

Comb honey

Frame for extraction

A comb for extraction (see photo 31) should always be shown in a bee-proof container with glass windows so that the exhibit can be seen by the show visitor. Timber for the container needs to be of a non-aromatic wood, such as hardwood: not pine, cedar or larch or any wood that gives aromas as this can taint both the honey's aroma and taste.

The judge will always test the comb for taste, ensuring that the frame is not sugar syrup. As a matter of interest, sugar syrup can normally be detected at the stage

of shining a torch through the comb as syrup always looks very dull compared to honey which is more reflective.

The comb should be fully sealed on both sides with an even surface and sitting proud of the frame. This allows an uncapping knife to slide down under the cappings without being fouled by the frame. It should have one type of honey and no cells filled with pollen, and also no sign of granulation in the honey or tunnelling from Braula.

Woodwork of the comb should be clean and free from brace comb and propolis. The comb should have no open cells and no staining from honey dribbling down the face of the comb. If after removing brace comb the frame is still wet with honey, you can put the frame back into a super and let the bees clean up the frame. Great care has to be taken as if you leave the frame there too long, there is a danger that the bees will

31 (above). Combs for extraction.
32 (below). A frame free of Braula.
Photos by Gerry Collins.

start to eat all the honey so I am talking of leaving the frame with the bees for less than an hour, more like minutes, not days.

Please be aware if your frame keeps dripping honey that it may be infected by Braula louse mite (not really a louse or mite but a wingless fly). This little beasty tunnels underneath honeycomb cappings with a 1mm-diameter tunnel up to 10cm long.

The insecticide strips used for varroa treatment virtually wiped out the incidence of braula and it was not seen at honey shows for a number of years. Braula infestation is now on the increase again, due it is thought to the treatment of varroa with organic acids rather than the insecticide strips we used in the early days of varroa infestation.

Further tunnelling and activity by braula can be stopped by putting the frame in a freezer. For show purposes the damage is done because the tunnels are still evident.

Woodwork of the comb should be clean and free from brace comb and propolis. A good tip for keeping comb woodwork clean is to use a little petroleum jelly: this stops the laying on of propolis by the bees.

Recap:
- Clean display case
- Display case lid that opens easily
- Clean frame free from propolis
- Good thick drawn frame with good yield
- Fully capped
- Wired if required for rotary extraction
- Same colour honey from the same flow
- No granulation if a liquid frame
- No debris on the floor of the display case

Cut comb

To harvest good show quality comb honey, you need a bit of luck on the weather front, both for normal floral honey and heather honey. Both need good consistent hot warm clammy weather to give a good honey flow.

It is advised not to attempt to produce comb honey until the OSR (oil seed rape) has finished flowering and the nectar from the rape has stopped flowing. The last thing you want, after going to all the trouble of setting up your hive for cut comb production, is ending up with a super full of granulated honey – notwithstanding there is a market for granulated cut comb and there are classes for granulated comb in some honey shows.

The hive needs to have a strong colony with supers set up with non-wired foundation (preferably drone), or starter strips. You, as the beekeeper, need to manipulate the spacing of the frames by swapping supers which have different spacing. To make life easy, use castellated spacers starting at 11 frames per super and swapping down to 10, 9 and even 8 frames per super. Sometimes it is worth doing this over a couple of seasons or flows, extracting the honey in between seasons by uncapping and letting the honey flow into a bucket, the frame being put onto wider castellated spacings.

Having carried out the preparations as mentioned above, you have encouraged your bees by manipulation of the super spacings to produce comb suitable for cutting, in that it is thick enough to achieve the weight within the capacity of the show container without adding liquid honey.

The process of cut comb, once the super is back at the honey room, is the same for both heather honey and other floral honeys (see photo 33).

33. Cut comb (Gerry Collins)

The first stage is to check that the combs you intend to use are:
- Capped both sides, rib is centre to the cut comb,
- No cells full of pollen,
- Not having been used for brood,
- Not granulated,
- No braula damage.

Drone comb: some people prefer drone comb, others prefer worker. It really does not matter. Obviously make sure if two boxes of cut comb are required for the class that both boxes of cut comb are the same e.g. drone or worker if showing a pair. It is prudent and good showmanship to cut the honeycomb the same way so that they look on the long dimension as they were on the frame.

The cutting can either be done with an off-the-shelf cutter or by making a plastic template of a size and of a dimension that enables a piece of comb that will just fit into the container to be cut.

Put the comb that has been cut out of the frame onto a sterilised kitchen cutting board. With a warm very sharp knife to cut the comb, make the cut following the lines of the template. Using this method rather than the off-the-shelf cutter gives the flexibility to trim, if your comb is overweight.

After cutting the comb carefully, place it on a cake cooling rack for draining, obviously putting the cake rack onto a plate so that all the drained honey is saved. By allowing the combs to drain, the resulting piece will be relatively dry. This gives a good impression for showing. Also, with the removal of all the running honey from the cut cells, the comb piece will not granulate so quickly. Pure heather honey should not granulate at all of course.

Check-weigh your cut comb piece with the container, but not necessarily putting the comb into the container at this stage. If the weight is to schedule and the piece fits into the container, you are ready to go.

Some show people freeze their cut comb until required for the show; others prefer to cut their comb a few days prior to the show so it looks nice and fresh.

Sections

To produce quality sections that win prizes at honey shows year in, year out, not only indicates a proficient show person, but also a very skilled beekeeper, because to produce sections you have to have both skills. There are very few skilled beekeepers in the UK who can produce sections time after time.

The old-fashioned method of producing sections was to throw a couple of swarms into a super with drawn comb. This achieved a very strong colony that was highly unlikely to swarm. The super was on a conventional floor; on top of the super was placed a queen excluder, and on top of that, a section rack: sometimes two or

34. Sections (Gerry Collins)

three section racks to give the bees more space. There was a chance in days of old when there were clover pastures, of the likelihood of filling more than one rack of sections. To stop the bees from just tunnelling up the racks, a piece of plastic cloth with a smaller dimension than the outside dimension of the rack was placed on top of each rack. A sheet of oilcloth was used; nowadays plastic is used. This technique pushes the bees to the outer edge of the supers or racks when they are travelling upwards to the next rack or super.

When the movable frame hive was first used for both commercial and amateur beekeeping, sections were the main method of producing comb honey. If you look at the old drawings of the original Langstroth hive, you can see the first section racks. The use of sections fell out of favour in the middle of the twentieth century during the Second World War.

The Ministry of Agriculture discouraged or even outlawed the production of comb honey during the Second World War, due to the wastefulness of comb honey production compared with extracted honey. We must remember that at this time there was a virtual Atlantic blockade and the population was not far off starving so any wastefulness was not allowed. The thinking behind this was that a lot of bee energy is spent producing wax comb. It must also be remembered that the Ministry of Agriculture were supplying denatured sugar to beekeepers for the bees' winter feed during the war, so beekeeping was very closely government-supervised, making sure the sugar went to the bees rather than in the jam pan.

Nowadays the main forage crops for sections are as follows:
- In arable parts of the country on the eastern side of the UK, borage is grown. This is a fantastic crop for producing sections; it has been recorded that bees, in very favourable weather conditions, will produce from foundation to a fully capped rack of sections in one week.
- In the dairy production areas on the western side of the country, we are seeing the resurgence of growing of white clover in grass leys. This is a fantastic crop for bees. In the past it was the main forage for bees in these areas. Recently there have been instances of beekeepers in grassland areas setting up their hives and leaving them in clover ley fields prior to moving to the heather with section racks, only to find the racks full and capped after a few days. Nice problem to have – with the happy task of rebuilding the section racks again ready to move to the heather.

The best show sections are produced when there is an exceptionally strong nectar flow, enabling sections to be capped very quickly without travel stains.

There are two types of section, round and square. These should be, and normally are, judged in separate classes.

The round sections are quite expensive to produce due to the cost of the hardware. However round sections are easier for the bees to work; like the rest of nature, bees do not like straight lines, e.g. squares. This is the reason that bees seem to like the round section racks.

The cheapest way to produce square sections is to use hanger frames that hold three sections per hanger. Using this method saves on the cost of buying the full racking system either for round or square sections.

Square sections normally weigh about 1lb when completely capped. Before showing you need to check-weigh the section or sections you intend to show against the schedule, making sure that the weight of your section conforms to the schedule requirement. A number of the premier honey shows schedule that you show a pair of sections. So there is a need to spend some time matching up your best pair of sections to be shown: never mix drone and worker.

Sections on the show bench will get marked down if the woodwork is covered in propolis. There are various methods to keep propolis off the basswood of the section, including paraffin wax, petroleum jelly, and various sticky tapes. One of the best is drawing-office tape: a very fine thin PVC tape that used to be common in drawing offices prior to the use of CAD. This tape does not leave a sticky deposit on the basswood frame when removed.

When setting up your hive for sections, you require very strong colonies, which will have a brood chamber boiling over with bees. One method for achieving this is to bleed workers from other hives in the apiary where the proposed section building colony resides:

- Pick a colony with this year's queen so there is no danger of swarming. The new queen is very necessary because the hive used for section production is going to be so congested with bees that they would otherwise swarm.
- Place the hive with the section rack next to one or two donor colonies, preferably in their own little apiary.
- In the middle of a warm day when the workers are all flying, remove the two donor colonies to another apiary. The remaining colony will take in the workers left behind from the removed colonies.
- After a further day, move the hives with the section racks to the required site or apiary.
- Never set up your hive using any supering other than section racks. Bees hate working sections so they will always work your other frames. If you are very confident of the honey flow and future weather, put on an extra section rack, with the plastic dividing sheet mentioned above between the section racks.
- For a further increase in density of bees in the hive that you are going to use for section production:
  - The hive is set up with brood box, on top of which is a queen excluder; on top of the queen excluder is a section rack.
  - Get two supers full of worker bees.
  - Place a piece of newspaper with slits torn into it on top of the section rack.
  - Leave for an afternoon and then overnight place a clearer board between the supers and the section rack.
  - The following morning, shake all the remaining bees out of the supers into the section racks.
  - The hive is then absolutely bursting with bees.

The above manipulation works really well for heather section production.

Processing the sections for showing and sale:

- Once the section rack is fully capped, clear the bees out of the rack and remove from the hive. The removal is done to alleviate travel stains on the sections (caused by bees walking over the wax cappings).
- If on removal of the sections from the rack you find that the basswood section frame is stained with propolis, this can be removed by using a Stanley knife blade.

- Check very carefully your sections for granulation and wildlife (dead wax moths, wasps and flies) and also check for other insect parts and debris.
- Make sure all cells are fully capped and there is no sign of braula, because if there is, you will end up with a sticky mess in your show box or sale carton. Braula-infested sections are not suitable for showing and really should not be sold.
- To show sections, use a glazed box which is bee-proof. These boxes are readily available from beekeeping suppliers and if kept clean can be re-used another year for another show.
- A good tip if you have a number of sections that you cannot sell or use is to cut the sections and use them for the Cut Comb class, as the depth of comb is normally just the right depth for cut comb cartons.
- Always show sections the right way up as worked by the bees.

Wax

Single wax block

Rather than give a specific method, the following offers a number of options which can be used depending upon the resources available to you. All the following methods will work with the various bits of equipment.

There are a number of sources for show wax:
- Wax cappings removed from your supers during the honey extraction. (Use an uncapping knife to get the best yield of wax.)
- Place the cappings on a mesh over your extractor or bucket to drain off the honey of the cappings.
- Nice new brace comb collected when doing hive inspections.
- Wild comb, collected normally when dealing with swarms, nice, new and not having been used for brood or stores. If the bees have used the comb for stores but not brood, place on a mesh as the cappings above.
- Do not be tempted to use wax for show that the bees have used for brood, or have used for stores for a number of years and odd bits of foundation.

All these sources of wax tend to produce very dark and dull wax.

Having got a source of wax suitable for showing, we need to wash the final bit of honey from the cappings. An old kitchen colander is ideal for this. If you live in a hard water area, it is advisable to use distilled water or to rinse your wax in rainwater from a very clean storage source.

Now we have to melt the wax and pre-filter to form a block. There are various methods to do this:
- If you have the option and use of a solar extractor, make use of it to melt your show wax. The solar extractor seems to help bleach the wax to a nice primrose

colour. With the solar extractor an initial filtering can be carried out using old tights or net curtains.
- Using a bain-marie on the kitchen cooker ring or hob, care must be taken not to boil the water vigorously or leave the melting wax unattended. It must be noted that melted beeswax is highly flammable and dangerous. When completely melted, pour the liquid wax over a piece of net curtain and run the molten wax carefully into the container mentioned later.
- Use a warming cabinet which has a thermostat set to less than 75°c: normally around 60°c for show wax as you just want the wax melted and able to flow. The higher the temperature that the wax is taken to, the darker it will go. Also, for safety, always have an override thermoset set to 90°c in case the first one fails. Some people use their kitchen oven satisfactorily but extreme care must be taken melting wax in the home cooker.
- With all the above methods when the wax is melted, pour it into a container lined either with greaseproof paper or a material that beeswax will not adhere to – an example being silicone.

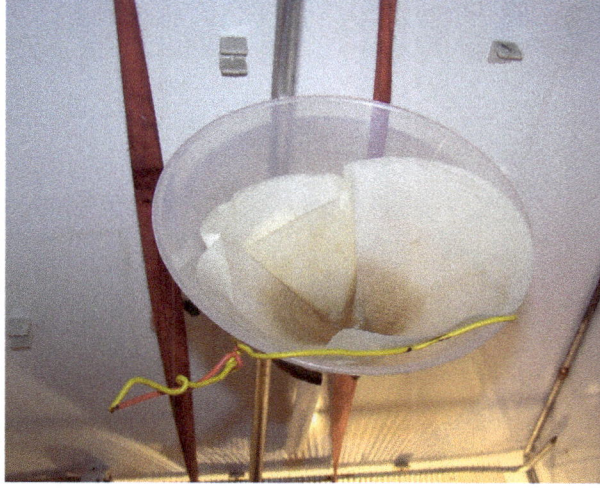

35. Whatman paper filters (John Goodwin)

Secondary filtering (normally reheating will be required) of the wax block:
- There are a number of filtering mediums available that have been used for filtering show wax: Whatman paper filters, (see photo 35), disposable J-cloths, coffee filters, nappy liners, tea bag material and surgical lint. There are many more that can be used. When choosing a filter material, make sure that the material cannot shed lint fibres or any other material as this ends up contaminating the showpiece.

The mould

Depending on the schedule, the mould has to be of a size to produce a cast wax block to the dimensions required. Most shows specify a required depth of the showpiece so the mould used has to have the dimensions and capacity to achieve this. To do a check for mould capacity, a good tip is to do a pre-weigh.
- As the specific gravity of wax and water is similar (wax is 0.963), water can be used to simulate beeswax to check the bowl capacity, by weighing the bowl on a top pan balance or similar type of weighing scales. Place the empty bowl on the scale, tare off the bowl weight, then fill the bowl to the desired weight, and mark off the water level in the bowl with a magic marker or similar indelible

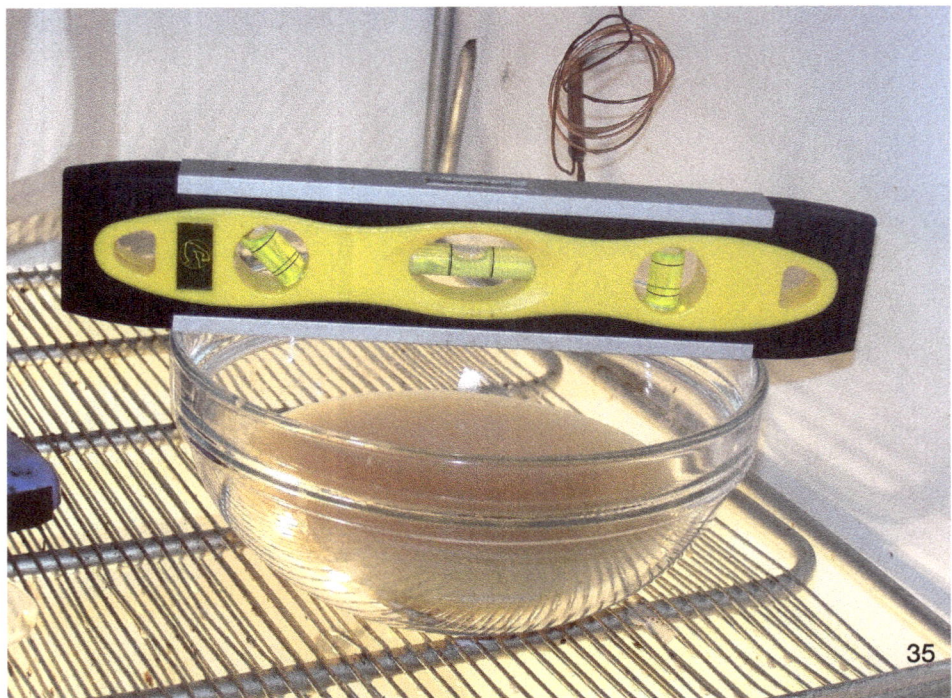

36. Checking the bowl is level (John Goodwin)

marker, having first checked the bowl is level using a spirit level (see photo 36).
- A glass kitchen mixing basin is ideal. Pyrex is really good, but make sure that there is no imprint on the bottom of the bowl. Unfortunately, Pyrex-branded basins now have the Pyrex imprint on the inside of the basin. All is not lost: if you search at your local large supermarket, their own-brand label basins made out of the same material as Pyrex (borosilicate glass) do not have an imprint.
- The basin or mould chosen should be used only for wax casting. General use will lead to scratches on the bowl surface. These scratches will result in a poorly finished beeswax showpiece as the surface scratches on the mould are transmitted onto its surface.
- A silicone mould can also be used; again, it is better just to use the mould for show purposes only.
- To prepare the glass mould, firstly we need to condition the mould with a release agent, that will release the wax showpiece when set and cold. In the past, green dishwashing detergent was used for mould conditioning. More recently, silicone release spray has been used.
- In the past silicone sprays used to leave an after smell of the spray in the beeswax showpiece. With modern sprays this problem seems to have been alleviated and they do not now leave an odour or smell. One of the big advantages of using the silicone spray is the resulting finish achieved on the beeswax showpiece. It is so good it virtually reflects the finish of the mould that has been used.

The secrets of showing

- The finish obtained by using green washing-up detergent is rather dull until buffed up with a polishing cloth. The reason that you cannot get a perfect finish with the detergent is that the detergent will break down some of the wax into soap, as the latter makes contact with the side of the mould via the very thin film of detergent. This soap film is readily buffed up with a polishing cloth but the finish is not as perfect as the silicone spray release method.
- If you decide to use the detergent method, a good way is to fill your kitchen sink with very hot water and add green liquid detergent until the water is green, then to submerge your mould in the water until completely covered. Leave the mould for a few minutes. Remove the bowl, being careful not to scald yourself and also being careful not to touch the inside of the bowl. Invert the bowl on the draining board and leave to dry. At this stage double-check the size of wax piece required for the show.
- If you have enough good quality beeswax and intend to exhibit at a number of shows, it is prudent, when all the equipment is available and set up, to cast a number of showpieces of different weights that you will require during the show season.
- Obviously, you will need a number of basins or moulds to carry out this production method.
- To produce a completely smooth piece of beeswax suitable for showing, the mould has to be at the same temperature as the molten beeswax. To enable the wax to cool evenly, we need to place a lid on top of the mould. Again the lid has to be preheated to the same temperature as the mould and wax: it can be a piece of very thick glass or an old thick dinner plate. Also, for good heat dissipation, warm a couple of house bricks wrapped in dustbin liners to place on the plate or glass lid.
- There are various ways of insulating the mould: e.g. by placing the mould in a water bath (the wet method). Again, the water bath has to be brought up to the same temperature as the mould and wax. Another method is to have the mould in an insulated container (dry method). If using a warming cabinet, the mould will not need any insulation.
- To enable the wax to cool evenly, we need to place a lid on top of the mould. Again, the lid has to be preheated to the same temperature as the mould and wax. The insulation and the heat-holding ability of the mould lid is the most important part of the wax-moulding process. As the other surfaces of the beeswax piece are in contact with the mould surface, the adhesion of the wax to the mould surface almost guarantees a good smooth finish. The top of the wax piece does not have this facility of adhesion to the mould side, so that the top surface of the beeswax piece is cooling at its own rate and not at the rate of the mould, leading to a rippled finish if the cooling rate is not controlled. So, the lid has to have the ability to provide a uniform surface of cooling by radiation as there is no other means of controlling the wax piece's top surface. Hence the use of a massive heat sink in form of two house bricks wrapped in a dustbin liner placed on the glass or ceramic lid or dinner plate, as mentioned above.

The process of pouring the wax begins with getting the mould level checked with a spirit level.
To hand, we will have:
- A warm mould with a fill level marked
- A warming cabinet up to temperature or an oven
- Enough melted wax
- Lid: dinner plate or a piece of plate glass
- Wrapped bricks for top of lid.

Pouring:
- Wax is poured as quickly as possible without causing a splash of the wax
- Lid is placed on the mould top
- Bricks (wrapped in plastic bag to stop dust contamination) are placed on top of lid
- Door of cabinet or oven is shut and sealed for at least 10 hours.
- Oven and warming cabinet are switched off

To remove the wax piece from the mould:
- fill the mould up with soft water and place in the fridge,
- or place the mould in a plastic bucket (always use a plastic bucket, not metal, to save the wax piece chipping) and carefully fill the bucket with cold water without damaging the cast piece. Put the bucket into a greenhouse, conservatory or similar area where the ambient temperature will increase and decrease during the day. The cast piece should then float to the top of the bucket.

Showing

When packing for the show double-check that you have packed the correct size and weight of block for the show you are about to attend. It is quite a common mistake for exhibitors to pack the wrong size of wax block for the show.

The wax block should always be shown in a sealed box or container that the judge can open for judging. See photo 5 on page 30 for typical showcases. A good tip is to line the floor of the showcase with dark blue velvet to show the wax off: the blue contrasts nicely with the primrose colour of the beeswax piece: see photo 37a right.

37a. Wax block (Gerry Collins)

Storage

The beeswax piece should be well wrapped in tissue or a material that will not scratch the surface of the wax but protect the piece from chipping. The wax piece should then be placed in a sealed container (biscuit tins are ideal). A good tip is to place your old filter medium or a piece of propolis in the sealed container taking care that these items do not come into physical contact with the beeswax showpiece. The filter medium and propolis give off aroma to the beeswax showpiece: a fresh smell as if it had just been cast. Never wrap or have cling film near to your beeswax block. Cling film will remove the finish on the wax block.

28-Gram blocks of beeswax

A similar process is used to source wax and prepare as for the single large block of beeswax. A number of shows request that you show six pieces, which is a little awkward as the standard mould from most suppliers is supplied with five moulding cavities. Thus, in preparing to show, two lots must be cast, either simultaneously using two moulds with five cavities each, or carrying out a second casting doing two castings of five.

As the moulds with cavities are made out of a thin sheet of rigid plastic, a good idea is to make a frame to fit round the moulds out of ¾-inch plywood. This gives the mould a bit of stability and some insulation. By also making up a lid of similar material, a mass of warm material can be placed onto the mould, helping more even cooling, as done with the large block of wax.

Mould treatment

To achieve the weight requested by the schedule, the same method is used as for the single beeswax block. To do a check for mould capacity, a good tip is to do a pre-weigh. As the specific gravities of wax and water are similar, water can be used to simulate beeswax to check the mould cavities' capacity. Firstly, check the mould is level, using a spirit level: this is highly critical in producing 28-gram beeswax pieces as any deviation from the level will affect the end weight of the beeswax piece. Using a top pan balance or similar type of weighing scales to weigh the mould, place the empty mould onto the scale and tare off the mould's weight. Then fill the first cavity to the desired weight and mark off the water level in the cavity with a magic marker or similar indelible marker. Then tare off the weight, read and repeat filling all five cavities.

To prepare the mould, use the methods discussed for casting the large beeswax block.

Moulding

To produce a completely smooth piece of beeswax suitable for showing, the mould has to be at the same temperature as the molten beeswax.

One way of controlling the cooling of the top surface is to put a massive heat sink in the form of a house brick wrapped in a dustbin liner on top of the lid. Having such a heavy lid on the plastic mould requires a wooden strengthening jig made out of a piece of ¾-inch plywood.

There are various ways of insulating the mould as we are having to strengthen it as mentioned above. The ¾ plywood cut to fit the mould acts as an insulator as well.

The process of pouring the wax starts with getting the mould level check with the spirit level.

To hand, we will have:
- A warm mould with a fill level marked in each cavity
- A warming cabinet or oven up to temperature
- Enough melted wax in a jug that pours well
- Lid
- Brick for top of lid.

Pouring:
- Wax is poured as quickly as possible into each of the mould cavities without causing a splash of the wax,
- Lid is placed on the mould top,
- Bricks (wrapped in plastic bag to stop dust contamination) are placed on top of lid,
- Door of cabinet or oven is sealed shut for at least 10 hours.

To remove the wax pieces from the mould, place the mould into a fridge, or in a plastic bucket (always use plastic not metal to prevent the wax piece chipping) and carefully fill the bucket with cold water without damaging the cast piece. Place the bucket into a greenhouse, conservatory or a similar area where the ambient temperature will increase and decrease during the day. The cast piece should then float to the top of the bucket.

Showing

The wax blocks should always be shown in a sealed box or container that the judge can open. For typical showcases see photo 5 on page 30. A good tip is to line the floor of the showcase with blue velvet to show off the wax, as the blue contrasts nicely with the primrose colour of the beeswax pieces: see photo 37a, page 80.

Storage

The beeswax pieces should be well wrapped in tissue, kitchen roll or a material that will not scratch the surface of the wax but protect the piece from chipping.

The wax pieces should then be placed in a sealed container (biscuit tins are ideal). A good tip is to place your old filter medium or a piece of propolis in the sealed container, taking care that these items do not come into physical contact with the beeswax showpieces. The aroma of propolis or the filter medium gives the beeswax showpiece a fresh smell as if it had just been cast.

Commercial wax block

This is a relatively new class, thought to have been first shown in Yorkshire.

A similar process to that for the single large block of beeswax is used to source and prepare the wax. However, a good tip is to use a block of beeswax that has been shown a few times and has lost its initial lustre: such a wax block is ideal for exhibiting in this class. Exhibitors will often use such a wax block to win in the commercial class. Another source of blocks for this class is a piece or block of wax that is not perfectly cast, as this class is judged on the colour, aroma and the wax's granular structure after the block has been broken by the judge.

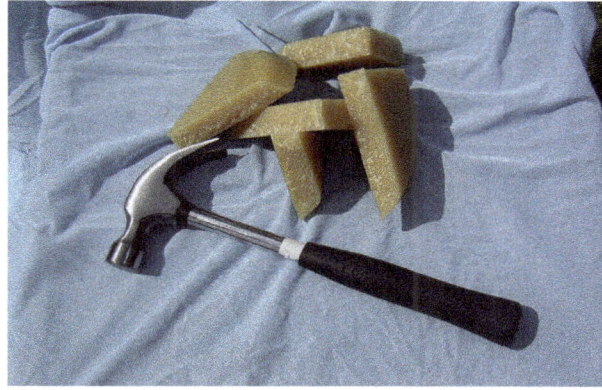

37b. Recasting (Gerry Collins)

The commercial beeswax class is one class where it is not possible to re-show the wax block, as it has to be broken to be judged. The only way that the block could be shown again would be to recast it: see photo 37b by Gerry Collins.

The problem with recasting a block of beeswax is that you lose some of the aroma, and the granular structure seems to degrade.

Make sure that the beeswax block you have available for this class conforms to the schedule. There will always be a minimum weight and normally a minimum thickness. Wax blocks should always be shown in a sealed box or container that the judge can open. A good tip is to line the floor of the showcase with blue velvet to show off the wax, as the blue contrasts nicely with the primrose colour of the beeswax pieces. As presentation of finish is not as important as with the other beeswax block exhibits, you can exhibit in a polythene container providing it is clean and does not give off any aroma which could affect the aroma of the beeswax block. It has been known for some exhibitors to show their beeswax block in a sandwich bag: needless to say, they did not win a prize.

Just as an aside, a Wax Chandler at the National Honey Show some years ago remarked that, as chandlers, they go to a great deal of trouble trying to eliminate aroma and colour from the wax, as the applications and industries market into which they sell their wax requires the product to be inert and white. The only application where an aroma and colour are required of the beeswax is church candles, for which they did not supply the wax.

So possibly, to be true to what is required for the commercial class, is not an exhibit with aroma or a nice primrose colour. However, read the schedule and make sure you know what the judges require.

Storage

The beeswax piece should be well wrapped in tissue, kitchen roll or a material that will not scratch the surface of the wax and protect the piece from chipping. The wax piece should then be placed in a sealed container (biscuit tins are ideal). A good tip is to place your old filter medium or a piece of propolis in the sealed container, taking care that these items do not come into physical contact with the beeswax showpieces. The aroma of propolis or the filter medium gives the beeswax showpiece a fresh smell as if it had just been cast.

Candles

Beekeepers often get asked to produce high-quality candles for churches of many denominations. Some of these candles – pillar and Paschal – are quite large in diameter and in height.

In the Roman Catholic Church rules in canon law are laid down for the blend of beeswax with other wax sources. This blend-ratio of beeswax to other waxes depends upon the service, celebration or devotion being held. For certain devotions, funerals, and during various periods of the church calendar (Lent and Advent) the beeswax used in the candles is not bleached and not blended.

37c. Beeswax candles (Gerry Collins)

Beeswax candles have been produced for hundreds of years; however, if you look back into the history of honey shows, it is relatively recently that beeswax candles have been scheduled to be shown at honey shows. This is quite extraordinary considering the history and the practical use of beeswax candles.

Candlemaking using beeswax is an art that can be practised and shown by non-beekeepers, providing of course that they have a ready supply of beeswax.

To show candles successfully, we must look at all the components of the candle and the method of production. Schedules will normally state a method of moulding and a method of production that is to be allowed and the required dimensions. Dimensions are normally stated as minimum and maximum height, length and diameter. If you are showing candles that are normally tapered, dipped, poured and cast using silicone moulds, measure the maximum diameter.

Tapered candles can cause a problem when showing because the wick size cannot cater for the complete profile of the varying diameter. If the wick is too thin for the candle's diameter, the wick cannot consume all the wax while burning and you end up with the candles pouring molten wax down the side. This called guttering. Guttering is very picturesque on Christmas cards but not acceptable for beeswax candle classes at a honey show. If the wick burns too fast you can end up with the wick tunnelling through the candle, spluttering and possibly extinguishing the light. Tunnelling is often caused by the wick not having been retarded enough to produce a burn consistent with the burn application, according to the type of wax used in the candle and the diameter of candle to be burnt. To retard the burn of the wick a non-organic material is blended into the wick, normally a salt solution. This process is called mordanting. (A very similar process is used to fix dyes in the textile industry: the same process name is used.) It is very important that wicks used are designed to be used with beeswax, as the burning rate of different waxes varies due to their having differing combustible characteristics.

Braided wicks: it was discovered that by braiding and putting one side of the braid under more tension, the wick would have a tendency to curl back over the candle on being ignited. Prior to the braided wick, the candle wick had to be trimmed as otherwise it would burn upwards, not burning hot enough to allow total combustion. This led to the burning wick breaking off the candle still burning and settling on the surrounding furniture, setting fire to its material. Braided wicks as used for candles nowadays were a massive revolution in increasing the safety of candlelit premises when invented – so much so that they were as big a revolution as the Edison tungsten light bulb was with the invention of electricity.

No matter what method of candle manufacturing you intend to use to enter a show, undertake test burns to make sure that the wick is suitable for the diameter of candle produced. Do not take it for granted that the purchased wick specification is correct for the diameter of candle you intend to produce. The best method for clarifying the correctness of wick to the diameter of candle to be cast is to carry out test burns. This is achieved by making or purchasing a conical candle mould. This can be burnt using the purchased candle wick to see at what point the wick is the correct burn for the diameter of the candle.

There are a number of differing moulding methods of beeswax candles for showing.

It is worth noting that some shows allow the use of wax from sources other than your own apiary. This is a very sensible rule as it encourages novice beekeepers to have a go at showing, even if they enter a rolled candle made out of foundation. (Of course, the exhibiting of commercial foundation would not be wax from the exhibitor's bees.)

It is a requirement that the wax used for candle showing is as clean and free from impurities as if showing a wax block. If the beeswax used for candle making is not refined and has particles of propolis in the wax, not only will you not win any prizes, but the candle will splutter and spit when burning. Propolis does not burn without semi-exploding – hence the spit and splutter. Other impurities within the wax can cause the candle to smoke. A smoking candle can sometimes indicate that there is contamination from other wax sources.

Refining and filtering the wax for beeswax candle production should be no different from the process required and used above for the production of beeswax blocks.

Poured

The pouring method is not very often seen at honey shows due to the complexity of the method.

For the poured candle process, a candle wick is pre-dipped in beeswax and formed so that the wick is still of a uniform diameter. It is then tensioned between two points. Liquid beeswax is then poured down the wick. It is also a good idea to set up a jig so that the wick can be spun round very slowly under tension, so that the wax has a greater surface area of adhesion to the already-wicked wax layer as it is poured. The wick is then un-tensioned and rolled on a glass surface by hand. To get a good result, this glass surface has to be kept at a temperature above ambient, normally about 25 to 30 degrees centigrade. During this process it is beneficial to have a very warm environment, completely draught-free with no fans circulating (even if a fan heater is part of the room's heating system). Movement of air will also move dust and this dust will get into the molten and semi-set wax, showing up as bits on the finished candle. This process of pouring wax down the wick is then repeated until the required thickness of candle is achieved.

If there is no method for keeping the wax plastic enough for rolling, the finish of the poured candle is very poor and not really suitable for showing. Using the heated rolling plate, the beeswax is kept in a plastic state long enough to take the required cylinder form by rolling. There are no known separate classes for poured candles at honey shows in the UK although the National Honey Show has a class for poured and dipped candles, where you do occasionally see poured candle exhibits. Unlike a dipped candle where there is always a consistent taper due to their production method, poured candles should have only a very slightly tapered profile.

We are discussing the showing of candles with the objective of winning the said class. Do not enter poured candles in a general class (where the schedule states any method of candle production) unless you are confident of producing a good piece of work and are confident that the judge will recognise it is a poured candle. For this reason, poured candles are hardly worth showing. However, please note an experienced judge will always mark up a poured candle over a dipped candle, a dipped candle over a moulded candle, a glass-moulded candle over a silicone-moulded candle, and a moulded candle over a foundation-rolled candle.

Commercial- and industrial-type operations still manufacture candles using a pouring or very similar process, using high-pressure extrusion technology. Long cylinders of wicked wax are produced and then are cut off to the length required and the top of the candle tapered to expose the wick, wrapped and sold.

Rolled

Although rolled foundation candles can look very smart, do not be tempted to show them in candle classes that allow other methods of production, such as poured, dipped and moulded. The only exception to this is if the foundation is home-produced. In this case, we have to be very careful to demonstrate a home-produced foundation on a rolled candle, as we rely on the judge to realise the method of production. As with all showing, don't take anything for granted: concentrate on the easiest path to winning. One class in which a rolled candle would be acceptable is the display class.

Moulding

For moulded candles, various mould materials are used – glass, silicone, stainless steel, latex rubber, polyurethane and polycarbonate.

Glass-moulded candles are very difficult to make compared with silicone-moulded ones. The finish and diameter tolerances of glass moulds are far superior to those of plastic moulds. The downside of glass moulds is the mould preparation required. Also, as the glass mould diameter is very accurate and constant, this results in efficiency of burning and wax consumption. The glass finish reflected in the candle mould finish is second to none.

As discussed under wax block processing above, a mould release agent must be used except with silicone and latex moulds, and to a lesser extent with polyurethane.

37d. Moulding process
(John Goodwin)

Moulding process (glass mould):
- To load the mould with the wick, firstly soak the wick in liquid melted beeswax. Allow for a good length of wick over and above the candle: it needs this extra length of wick to help extract it from the mould.
- Stretch and hang the piece of wick and allow it to cool.
- When the wick has been threaded through the bottom of the mould (which will become the top of the candle), it then has to be sealed with Blu Tack or a similar material.
- To centre the wick and keep it in tension, see photo 37d. These simple home-made devices are made out of the bottom bars of old hive frames cut to

- length; the two pieces are then held together with elastic bands.
- To help achieve a good finish, it is advisable to insulate your mould: this can be done using plumbers' pipe insulation – see photo 37d.
- If using glass moulds, they have to be to be treated with a mould release agent prior to filling. Two options as release agents are washing-up liquid or a silicone release spray. If using the washing-up liquid, the method is to submerge the moulds in a strong solution of very hot water, remove the mould and drain until completely dry. If you use a silicone release spray, make sure that the spray covers all the mould surface that is in contact with the wax candle. One method is to spray a piece of kitchen roll with the release spray. Wrap the sprayed kitchen roll piece round a small kitchen fork, then push the fork down to the bottom of the mould and rotate. If using silicone moulds, a release agent is not really necessary.
- To help keep the moulds vertical, a good plan is to tie a number of moulds together, giving the moulds a bit more vertical stability and placing all the moulds in a large container. See photo 37d using a laboratory beaker.
- Filling the mould needs to be done as quickly as possible, allowing plenty of wax. A good plan is to cast a number of candles at the same time. Not only does this make the casting process efficient, but it enables all the wax to be of the same batch, so that the colour and aroma will all be the same. This is very important for showing when there is a class requirement for multiple numbers of candles to be shown.
- When all the moulds to cast have been filled, have a quick check to see if any moulds need topping up and also check that the wick is still centre of the mould.
- Shut the warming cabinet door and leave overnight or for a number of hours.
- Place moulds in the freezer for a day.
- Remove from the freezer and pull out by applying pull pressure to the wick.
- Trim the bottom of the candle with a sharp knife so that the bottom is flat, checking that the wick is central to the candle.
- Trim the top of the candle wick and then very carefully dip quickly into some molten beeswax to give your wick that final finish.

Fettling and polishing your candles

- Before polishing, check the schedule to see if polishing is allowed. If it is not mentioned in the schedule rules, carry on.
- Remove all burrs with a very sharp knife, being careful not to chip the candle edges.
- Polishing your candle or wax pieces can be done using surgical spirit (isopropyl or ethanol) which can be purchased from any high street pharmacist.
- Safety warning: some people have a skin reaction to this substance. If so, wear hand protection.
- Before using the above substance, it is best to try a little on a piece of wax which is of no consequence if spoilt.
- Apply by putting a little surgical spirit onto a piece of silk (old underwear is ideal).

- Rub the material gently on the candle in one direction: the finish will be outstanding and like glass. In effect the finish is virtually French polish.

Silicone moulds and other soft plastic moulds

- Using silicone moulds is a lot simpler than glass moulds: you do not have to prepare the mould other than wick it.
- It is not necessary to pre-wax the wick, but if pre-waxing is carried out it saves dipping the wick after moulding.
- To wick the mould, a good tool is an old-fashioned darning needle which helps push the wick through the hole in the mould.
- The wax used for silicone moulds does not have to be as hot as for glass moulds, but care must be taken to meld the wax enough. If the wax is too cold, the moulded candle ends up with bubbles on the surface of the wax.

Dipping candles: the process

- Cut the wicks to suit the diameter of candle. If for showing, do not calculate on the maximum diameter. At a show, the maximum diameter will not be achieved even after quite long double burns. Candles are lit twice while judging: the second ignition is to make sure the candle will relight.
- If you intend to make your dipped in candles in pairs, a good tip is to tie a thread nut to each end of the wick, immerse in the dipping tank, then remove putting the wick under tension until it has cooled. You can then remove the nuts as the wick is stiff enough to carry on dipping as usual. A candle will need to be dipped between 10 and 20 times.
- The length you cut your wick to will depend on your dipping rig or racks and method. Nowadays, exhibitors seem to use racks which automatically tension the wick, so the length of wick cut should at least be a third longer than the length of the desired candle.
- The kit you will need for dipping is a hot plate or a warming cabinet to keep your wax warming, and a dipping tank that is 100mm to 150mm longer than the length of candle being produced and of a diameter to allow submersion of the rack you are going to use.
- Dip quick, cool long. So dipping should be carried out quickly, and then the rack removed and allowed to cool till the wax is very plastic. Carry on dipping until the desired diameter is achieved.
- Rolling: if possible, have a piece of plate-glass kept at a temperature above ambient. This gives more time for rolling while keeping the wax in a plastic state.
- The ambient conditions need to be draught- and dust-free. If the candles are in a draught, they will not end up perfectly round. If you have dust in the atmosphere, this dust will get onto the surface of the candle.
- Polish the candle as described for the glass-moulded candle, if required.
- Keep the candles in a sealed tin with a lump of propolis or the dirty filter used to filter the wax. The type of tin malt whiskey is sold in is suitable: they keep

the candles upright. Dipped candles seem to bruise very easily, and if they are kept upright, they cannot bruise along their length. Polishing the candle does harden the surface somewhat.

When transferring your candles to the show, wrap them in soft tissue paper. Don't use any woven material as the fibres sometimes get stuck in the candle or showpiece. (See photo 37e).

When showing your candles, make sure whether the show management are supplying candleholders, and if you supply the holders, that they are fireproof. Also check your schedule to see whether the show management are going to use spikes as candleholders. If spikes are to be used at the show, remember that not only will the lit candle not be usable for another show but neither will the unlit candle which will be spoilt.

37e. Showpiece (Gerry Collins)

Wax models/flowers

This class is all about artistic perception. The two main forms are cast and fabricated. Wax flowers are normally constructed using thin wax sheet; curvature of a piece can be achieved using various sizes of spoons. When dyeing wax for a particular colour, always remember that the base wax is yellow.

For stems, use florist's wire and tape. There are two main types of tape, one more realistic but harder to use. A good tip is to split a tape lengthways for a better-looking result and to save tape.

Wax casting of fruits and vegetables requires the making of your own plaster cast, using quality plaster suitable for casting models. See photo 38.

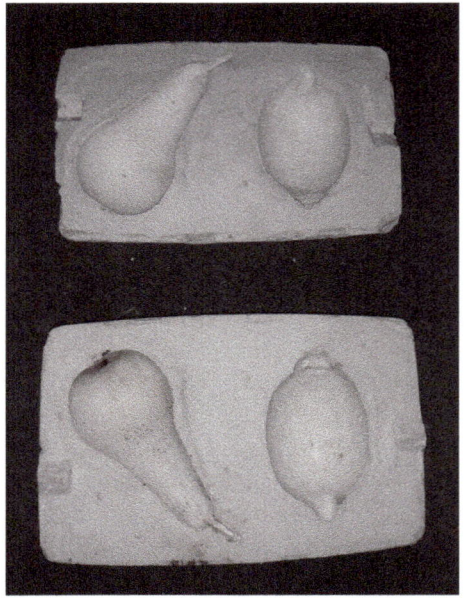

38. Casting models (Joyce Nisbet)

Beeswax polish

Polish is normally shown in polish tins, with no sales label as with other exhibits in a honey show where the exhibitor is anonymous. If the polish is shown as for sale, there needs to be a hazard warning label (CLP Regulation) and the exhibitor must read the necessary regulations for the sale of polish. CLP means Classification, Labelling and Packaging of substances and mixtures. The label should be attached or stuck to one of the container's surfaces.

Prior to producing your polish, read the schedule. If the schedule specifies a cream polish, you need to add soap flakes to the following recipe.

A recipe used for producing polish for showing:

- Place in a Pyrex basin equal parts of natural turpentine and broken bits of beeswax, the smaller the better. An old cheese grater is useful for this process, or pieces from previously-shown commercial wax block or a candle piece that has been lit are ideal. Make sure that there is absolutely no trace of ash or black bits.
- The Pyrex basin used obviously should not be used afterwards in food preparation, nor should the basin be used as mould for wax blocks. Beeswax polish taint seems to stick to moulds even after thorough washing. Perhaps this is why it is widely used for furniture polish because of the lingering aroma.
- Safety warning: making polish is very dangerous if care is not taken. Do not under any circumstances leave the mix on the stove unattended even for a minute.
  - Place the basin on top of a pan of warm water.
  - Carry on warming the mix using a bain-marie technique or double boiler.
  - Open enough polish tins to have the capacity to take all of the mix being melted.
  - Make sure that the polish tins are on a surface covered with sheets of newspaper that can be disposed of if there happens to be spillage.
  - Check with a spirit level that all the tins are level (we are producing to show, remember).
  - When the wax has completely melted into the turpentine, pour the mix into the polish tins.
  - Wait until cool before putting lids onto the tins. Check that there are no bubbles on the surface of the polish.
  - A tip is to make your beeswax polish when there are no other show items around in the vicinity of the polish. Putting it bluntly, beeswax polish stinks and can taint other exhibits like wax and honey.
  - Showing your polish: make sure there are no rust spots on the polish tin and that the polish has not dried out to leave a crack around the wax.

Display classes

Victorian pyramid

"Victorian pyramid" is a display usually made with plate-glass shelves and glass columns decreasing in size to form a pyramid (see photo 39). The schedule will have a maximum footprint size allowed (about a metre square) and sometimes a height limitation.

With regard to display items it is quite usual to have:
- honey of any variety, any colour, comb (both cut and full frames), granulated, liquid, chunk and heather;
- wax in any of the following forms: candles, blocks, statues, flowers, fruit, coloured or natural;
- mead shown in exquisite containers: mazers, silver-lidded jugs and ewers.

39. A Victorian pyramid (Joyce Nisbet)

Read the schedule carefully to ascertain minimum amounts of produce required.

This class is one which stewards do not stage. Staging is always done by the exhibitor, as a large part of the marks are for the artistic display

Method of showing
This is one class where you cannot be disqualified on the quality of the produce. The only disqualification could be on the sizing of the display and the weight of produce provided.
- Take time to build the display prior to transporting to the show.
- Photograph the display before dismantling.
- Check each jar of honey as you would for all the honey classes. See notes above under: Extracted Honey (Dark, Medium, Light), Granulated Honey, Soft Set, Chunk, and Heather Honey.
- Needless to say, the props, shelving and support devices should all be highly polished and very clean.
- Go for surfaces that reflect light as much as possible. Plate glass is ideal.
- Check each piece of wax: see notes above. Blocks, candles, small blocks, wax statues, wax fruits and wax flowers are often set up as small floral displays or as floral vines rambling up the columns and onto the glass shelves. Wax should be displayed in sealed containers.

The secrets of showing

- Mead shown in artistic containers: lidded jugs, mazers and ewers. Mead has even been shown in the columns supporting the glass plate shelves, starting with a dark-coloured mead at the bottom of the display with a light-coloured mead at the top.
- The same colour profile as the mead could be matched by varying-coloured honey going from the bottom shelf with dark honey through to light honey on the top shelf.
- Products of the hive are sometimes allowed in the schedule, but if you decide to show products, only do so if they are of artistic merit.
- If you are showing fancy moulded wax candles, a good tip is just to cast without the wick.
- If showing a full frame of honey, make sure that the display case is spotlessly clean or use one made out of stainless steel.
- A very good container now available for cut comb is the new crystal comb container made out of clear plastic. This allows the light to shine through the comb honey via the container.
- Jarred honey: although advice was given in the Extracted Honey section above to use plastic lids, always use gold metal lids for the display class, for their reflection. Check very carefully for any signs of rust (often found on the turnover near the end of the thread). It does not matter for the display class if the gold plating is of a different hue on the jar lids as they are not being judged as pairs.
- The objective with this class is to have a wow factor. If every item is looked at by the exhibitor for its reflective effect, this helps the display's wow factor.
- Don't have too many items boxed in cardboard. This absorbs all the light and gives the display of sections a dull feeling. If you are exhibiting sections, purchase coloured boxes. It is all a matter of balance: if you have very good quality sections, by all means show them but bear this advice in mind and don't have boxes stacked upon boxes. Never show sections laid down on their side.
- Make sure that the size and stock amount of produce conforms to the schedule.
- Recheck the photo against the display set-up.

Shop window display

Follow the steps above as for the pyramid display. However, as this is a shop window normally made out of three Georgian-type unglazed wooden window frames, there is not the reflective finish of the pyramid display. The display schedule will specify minimum amounts of produce to be exhibited, a size limitation and very often a tighter rein on what can be shown. As this is a shop window display, sales regulations should apply, but check the schedule.

40. Shop window display (Gerry Collins)

Mead

Even small honey shows will have at least two mead classes: dry and sweet. Scheduling normally states what type of bottles the mead must be exhibited in (normally 70/75cl Bordeaux style and must be punted). As with all other exhibits there should be no stamped lettering on the bottle and it should be of clear glass. Cork stoppers with white flanges have to be used. A good idea is to change your cork prior to the show so that at least you will get top marks for the condition of the cork.

Fill levels should be to 15mm between the top of the liquid and the bottom of the cork.

Mead shown should have no other alcohol added. The only additions that are allowed are tannin, nutrients and food acids.

As with other show entries, the only labelling allowed is as supplied by the show organisation or by an instruction as to label type.

There are many recipes and methods for producing mead: certainly too many to mention in this book. The following notes are really an addition to the recipes provided in normal mead-making books. All the notes are really geared to help win prizes and not to just produce a mead suitable for home consumption.

Type of mead

- **Pyment**: red or white grape. Obviously, the source of the grape juice will affect the final colour of the mead.
- **Hippocras**: cinnamon. It should be noted that the name Hippocras is used in some eastern European countries to describe grape wine with cinnamon. In some countries it is sometimes a mixture of pyment and cinnamon.
- **Cyser**: apple or crab apples (if crab apples are used not much tannin is required) and normally quite dry. Just as an observation and certainly not in agreement, sweet cyser is sometimes marked down against a dry cyser.
- **Metheglin**: spices. Metheglin is thought to be where the word "medicine" comes from. The blend of spices has to be used with care as it can end up tasting like cough medicine in one extreme and like liquid Christmas pudding at the other.
- **Melomel**: fruit and fruit flowers (elderflower being an example) but not apples or the juice of grapes.
- **Hydromel**: normally watered-down or low alcohol; or even no alcohol.
- **Sack**: fortified, it is one of the stronger blends often with unfermented honey added. Normally very sweet.
- **Bochet**: burned or caramelised honey. Ideal for making use of overwarmed honey from the warming cabinet. Show with care as it is very seldom shown so the judge may not recognise it as bochet.
- **Bochetomel**: a bochet-style mead that also contains fruit such as elderberries, blackberries, raspberries and blackberries.

The secrets of showing

41. A variety of meads (Gerry Collins)

- **Bragot**: also called bracket or brackett. Originally brewed with honey and hops, later with honey and malt, with or without hops added. Welsh origin (bragawd).
- **Capsicumel**: made with addition of chilli peppers (USA derivative).

To make mead for showing, here are three pieces of advice: 1. be patient; 2. use the best ingredients and 3. use a good recommended recipe when starting out on your mead-making experience for showing. It is not the time to start experimenting with your own recipe until you have success on the show bench.

It is possible to produce very good meads and even show-winning meads from honey that is not the best. However, as we are discussing the showing of meads, to brew meads with a better chance of results and winning prize cards, use the best-tasting honey you can afford to spare to make your mead with.

Basic mead is just water, honey and yeast. These three ingredients make a very basic mead that is not particularly good tasting. To give the mead some bite and balance the following ingredients are what make the difference

Acids

Food/ fruit acids are added to the mead. If acids are not added, the mead will be just a very aromatic sweet alcoholic liquid, very bland and flabby in taste. Acids help balance the taste of the mead.

Yeast needs an acid broth. Honey is not acidic enough to produce a good fermentation.

Acid also helps control unwanted bacteria in the must (the name given to the mix by brewers and winemakers prior to fermentation).

Acid sources are citric, tartaric, and malic acid.

If you are making a melomel, cyser, pyment or other combination, some of the acids will not be required.

Water

Living in a soft water area, tap water is adequate without using other water sources. Mead-makers living in hard water areas sometimes use distilled water, or rainwater (care must be taken in using untreated water sources). Do not underestimate the amount of care that goes into the treatment of water supplied to us by the various water and utility companies. In fact, as an aside, there is strong evidence that the first water treatment method was making mead, the fermentation process killing most of the bacteria in the water to be drunk.

Honey

A lot of literature suggests that honey washings, baker's honey, fermented honey and honeydew can be used for brewing mead successfully. However, for showing mead, the best honey that has been harvested and can be afforded should be used.

All types of honey are suitable for making mead, from the very light and mild tasting to heather honey and dark honeys. Some dark honeys can give a unique flavour, which is fine for home consumption but be careful of showing such mead.

Honeys such as brassica and legume honey (oil seed rape [OSR] and clover) are very light-coloured, so will produce a light-coloured mead of a primrose hue. This is a colour sought after for showing, and is normally shown and produced as a dry mead.

The heather and darker honeys can produce honeys the colour of vintage port. A little care has to be taken using heather honey as the moisture content is normally somewhat higher than normal floral honeys. The difference can be up to 5% so allowance in the recipe has to be made when using heather honey.

Yeast

The yeast that is required depends on the type of mead you are brewing: there are a lot of different brands. The Gervin range is very popular. It is produced by Muntons who give a description of the application for each of their yeasts, rather relying on a wine type description: Chablis, Sauterne and Burgundy. Trying to match a mead type to a grape wine type is not easy. You take your choice whether to use liquid or powdered yeast.

Tannin

A lot of recipes suggest the use of a cup of strong tea added to the must. If one is producing melomels or other fruit meads, adding a large amount of tannin is not recommended as the skins of elderberries, apples, plums, and grapes contain tannin. Too much tannin will unbalance the mead making it very astringent.

Vitamins and supplements

Honey has not got a complete nutrient profile for yeast so other nutrients and vitamins have to be added to the must:

- Vitamin B1 (thiamine) helps the yeast break down carbohydrates.
- Vitamin B3 helps the yeast convert fats, carbohydrates and proteins into energy that the yeast can utilise efficiently.
- Vitamin C is an anti-oxidant and helps prevent wine spoilage.
- Necessary nutrients in the form of nitrogen, phosphates
- Magnesium sulphate helps the yeast perform in the mead-making environment.

It is strongly recommended to follow a good recipe to start your mead-making journey.

Equipment needed
- Plastic bucket (food grade)
- Jam pan (stainless steel)
- Filter or sieve food grade plastic or stainless steel
- Demijohn
- Funnel
- Airlock
- Plastic pipe
- Bottle brush
- Hydrometer: see description below
- Jam spoon, wooden or hard plastic
- Laboratory rubber bung (with a diameter to fit in the neck of the demijohn) with air lock
- Wine bottles and corks to suit the schedule and specification for the requirements of the show you intend to exhibit at.

To measure the alcohol content of mead there are two instruments that can be used. Both use an indirect method. A hydrometer measures the buoyancy of the liquid (mead). The other solution is a refractometer which requires the use of an alcohol table, and measures the liquid (mead)'s refractive index. Both these devices are within the economic reach of a mead-maker. In both cases there is a more expensive digital solution.

A hydrometer is as important a piece of kit as all the other utensils listed. Most mead and wine makers use a hydrometer that has been calibrated for the specific gravity within the range of meads and wines (1.00-1.110).

To achieve a really accurate measurement, a laboratory cylinder should be used and the liquid being measured should be at 20°C, as the calibrated scale on the hydrometer was calibrated at this temperature. The cylinder should be filled to within 40mm of the top. Care should be taken when reading the result: the reading is taken at the bottom of the meniscus curve (see photo 42). A gentle turn of the cylinder should be taken to make sure that there is no hysteresis (stiction) between the cylinder and the hydrometer glass surface.

This reading is a guide to the potential alcohol content of the mead. A really accurate measurement can be gained by calculating the alcohol by volume (ABV). This is calculated by measuring the start gravity prior to fermentation to the finished fermented mead, counting the amount of gravity drop and dividing by 0.7362. A simple method is to take both readings and use the ABV calculator on the web. If there are additions of honey during the fermentation process, the hydrometer reading will have to be retaken.

42. Laboratory cylinder (Gerry Collins)

The difference between sweet and dry mead is of course subject to taste. However, a good guideline for showing mead would be any specific gravity of 1.005 to 1.009 and a really sweet sack/dessert type mead could have a specific gravity of 1.020+. A dry mead would normally have a specific gravity of 0.099–1.006.

Cold pre-fermentation process

This is kinder to the ingredients. There are sometimes issues in getting the mead to finally clear and some say some of the aromatic flavours lost by the boiling or hot method are preserved with this cold method. To counterbalance this idea of a pure method, sometimes recipes and mead-makers add a Camden tablet to sterilise the must prior to adding the starter yeast. There is always the risk of making vinegar rather than mead if Camden tablets or sodium metabisulfite are not added in some form. Camden tablets can sometimes slow or check the fermentation process.

Hot water process

The hot or boiling method requires that the must, in a concentrated form (cold sterilised water is normally added later), and other ingredients (fruits and spices) are boiled, covered and then cooled prior to the addition of the yeast to start the fermentation process. Again, with this process there are arguments for and against: boiling can degrade the must, and can also be very time-consuming with two boiling processes to carry out, as the fruit additions are boiled separately. One argument put forward is that a lot of the protein in the mix not required by the yeast is removed on the first skim or filter as the particles have a tendency to coagulate or attract each other and become quite large, so are easily filtered or skimmed.

Mead-making is really a balance of hygiene, sterilisation and not over-processing the ingredients. Ideal mixing of the must is at about 20°C when the must has been charged into the demijohn for the fermentation to start. The ambient temperature needs to be below 15°C: this steadies down the fermentation.

Two items that you do not require when making mead are children and pets. For some reason, the pollution and items found when judging mead is amazing – items like dog or cat hairs and other animal parts. Children don't leave parts, just fingerprints.

Honey beer

The art of using honey in brewing is to try and achieve a taste of honey but not to kill completely the bitterness of the hops and the beer taste, as you will end up with a beer tasting of honey like mead (Braggot). It is not intended to give a recipe for the honey beer. This is left to the exhibitor to try different recipes or one specified in the show schedule.

Kit required
- Brewing pan
- Carboy or large demijohn with a large airlock
- Laboratory food-grade plastic funnel
- Plastic or wooden spoon
- Sanitiser: if using plastic, be careful of chlorine as it can leave an aftertaste.
- Plastic tube to be used for syphoning
- Bottles
- Crown caps and applicator
- All the necessary consumables to make your brew

When making honey beer, there are three methods: one is to add the honey to the initial steeping and boiling. This option controls and kills any wild yeast in the honey: the disadvantage is that you also kill the honey's subtle aroma and taste.

The second method is to make up a solution of honey and hot water, keeping it at about 80°C for a period of time (20 mins. minimum). This seems to check the wild yeast but not spoil the delicate honey aromas, so cool and seal it with a cloth cover and add to the cold wort.

A third method is to put some honey in when bottling. The secondary fermentation gives the best flavour but you can end up with cloudy beer.

Confectionery

Confectionery classes cover a multitude of classes from cakes, biscuits and sweets to jams, curds and chutneys.

Most failures in this section are either from using the wrong container (baking tin) or not cooking correctly. There is no better class in this section than a plain honey cake as you get the aroma of the honey. Some classes have such a lot of fruit and strong-tasting items like chocolate that the aroma of the honey is completely lost. As with all the other sections and classes, make sure that the schedule is understood and followed.

43. Honey cakes on show (Gerry Collins)

Showing

This is one section with classes where the exhibitor is usually allowed to use honey from sources other than their own apiary. This is to encourage novice beekeepers or non-beekeepers to exhibit.

If the schedule allows, and if the show management do not provide them, a good way of showing off your exhibit is in a plastic cake dome. Cakes should always be shown in sealed containers (we are showing a food product). If you have not got access to a cake dome, a clean sealed plastic bag could be used.

Read the schedule carefully as sometimes the schedule is not specific about the honey type to be used in the recipe. If the type of honey is not specified, a good tip is to use a little strong honey, which is very aromatic compared with a light honey. Care has to be taken using heather honey for cooking, as it can very easily help burn the cake. When heather honey is used, the aroma is second to none when the cake is cut.

One of the problems of baking with honey is that it is a lot more acid than sucrose, so most recipes allow for bicarbonate to help neutralise the honey. Another problem with honey baking is that the sugars in the honey react differently from sucrose. If you are baking a yeast product, the glucose in the honey will immediately feed the yeast leaving the fructose: this can sometimes be indicated in the bake.

If you are baking a fruitcake, make sure that all the fruit required is added. Do not try to substitute as the judge always goes to the trouble of checking the fruit. Also be aware that some fruits do not look as one expects: a good example is organic cherries as they are a completely different colour from the glossy reds that you expect them to be. If the judge is not aware of this, you could be marked down. Always consider that the judge may not know the difference.

Be aware that the judge will always taste and cut your cake and leave the cut to be seen by the public.

Jams and preserves: check the schedule very carefully. Some schedules require standard lids; other schedules require cellophane and wax discs.

Hives: Observation hives

There are a number of different types of observation hives, from the nucleus type, where you just see one frame of bees, to the more traditional observation hive constructed as if it was a section of a double brood colony cut from the hive. With this hive you get two brood frames stacked, a queen excluder, and then one or two super frames stacked again vertically on top of the queen excluder. (See photo 44 overleaf).

To set up an observation hive, read the schedule very carefully. The observation and nucleus hive classes are not usually open classes due to the requirements of biosecurity. Entries are normally limited to the local association members or members of the public living in the show's local geographical area.

Setting up your observation hive

- Make sure that the observation hive is absolutely spotless prior to using and populating: woodwork nicely varnished, the glass armoured or a plastic

44. Observation hives (Enid Brown)

type so that it is not easily broken and can't shatter. Make sure that the glass windows shine brightly.
- Ensure there are watering and feeding facilities or ports on the hive. Read the schedule: some show schedules insist that water and sugar syrup are available at all times. If this is stated on the schedule, make sure that you as an exhibitor provide squirting bottles or containers. If you are allowed to, keep an eye on your bees at the show. Legally the show is responsible for your bees while the show is running. However, morally you are the custodian of your bees.
- To be very pedantic, make sure that there is no sign of disease. If you have any doubt or there is any sign of disease, under no circumstances take your hive to the show. You are legally required to inform your local bee inspector if you are concerned that the bees have a reportable disease.
- Plan which hives and which frames are going to be utilised long before the actual setting up of the observation hive.
- To have the best frames for showing both brood and super frames, use frames that are as new as possible but have held stores and been extracted. This allows the bees to start off with a completely flat drawn-out comb. This is best achieved if you put brood frames above your queen excluder letting the bees draw out the frames, fill with stores and then you extract, leaving very level and fully drawn-out frames. The resulting frames are drawn out to the bottom bar of the frame.

- If you configure your donor hive as a double brood box or a double nucleus hive, you will have the benefit of using broad frames in your observation hive giving a circular brood nest: the top brood box will have pollen and stores above the brood, whereas the bottom brood box frame will just have brood to the top bar. Put into the observation hive this will provide a circle of brood.
- The best observation hives for showing are ones that represent a vertical cross-section of the hive, as mentioned above, so that the public see a fully circular brood nest. Above the brood nest you have the queen excluder and a frame of stores not completely sealed so the public and judge can see the full cycle of honey production.
- Use only very clean drawn comb with clean frames.
- Mark or re-mark the queen just at the time of the transfer to the observation hive. If the schedule specifies, mark the queen as the birth year, or with a white mark.
- When the hive is completely set up, place on cover boards ready for transport.
- Fill the hive well because if the bees are free-flying, they will regulate themselves at a density for the queen to be seen. The normal problem with most observation hives shown is that the hive population is far too low. This is fine for the public but not for the bees as they need a reasonable density to keep warm. The judge will mark down more for a low-density colony than for a high-density colony.
- The entrance tube – good modifications to fit to it are:
  - a Porter bee escape that can be mounted as a slide valve so that when the time comes to shut up the hive when the show has finished, the Porter bee escape is slid into the entrance tube allowing bees to return but not exit the observation hive. (See photo 45).
  - a piece of the entrance tube made of Plexiglass so that the public can see the bees travelling along the tube. Along with this modification, there must be a method of covering the tube so that most of the time the entrance tube is in complete darkness. If the tube is not in darkness the bees become confused as they push onto the wall of the tube trying to follow the light source.
- Mark up the hive as per the schedule. If the labelling is not specified, a good plan is to use small numbered labels and then have a key with a plastic poster giving a more detailed explanation of the hive community and structure.

45. Porter bee escape (Frank Kirkham)

Although observation hives attract the public and have a positive effect on beekeeping public relations with the general public, the utmost care should be taken when showing observation hives. It is not unknown for the colony to be lost due to heat at shows. Incidents like this negate all the PR that a good observation hive achieves.

There are a number of different types of observation hive. The traditional observation hive is basically a cross-section of a double brood box hive. This is the type of hive that is suitable for multi-day competitive showing with free-flying bees. The small nucleus type observation hive where only one brood frame is visible is only suitable for school demonstrations and is not really suitable for competitive showing.

Hives: Nucleus hives

At the start of the season when splitting and making up nucleus hives, pick your best nucleus hive using a new nucleus brood and a very clean or new set of hardware in good condition. Read the schedule and confirm the size and number of frames. Be aware that the show management may require your bees to be used for demonstration purposes. Obviously, the bees for showing must be of a very calm and easily handled nature, not only for winning or being placed in class, but so that the bees can be demonstrated.

- Pick your best nucleus box hardware: brood box, lid, cover board and travelling frame. Make sure that all components are spotlessly clean and put this box with all its parts and bits on one side until nearer the show date.
- To have the best brood and super frames for showing, use frames that are as new as possible but have held stores and been extracted. This allows the bees to start off with a completely flat drawn-out comb. This is best achieved if you put brood frames above your queen excluder, letting the bees draw out the frames, fill with stores and then you extract leaving very level and fully drawn-out frames. The resulting frames are drawn out to the bottom bar of the frame.
- Out of the nucleus colonies you have been running, pick the one with best brood pattern, nice temperament, and brood at all stages.
- Find the frames with the best-looking stores you have, again on nice clean evenly drawn-out frames, and amalgamate in a full-size brood box.
- Nearer the show date you then have the chance to pick the best of all the frames you need for the show.
- Transfer the number of frames according to schedule to the reserved nucleus box.
- Check the queen marking and make good if it is not very clear. This needs to be carried out at least a week before the show. The last thing you want is for the colony to ball the queen prior to the show.

Handicrafts

This title covers such a lot of different skills including needlecraft, pottery, sculpture and pictorial art. Read the schedule to make sure that you conform to the basic requirements of dimensions and art medium. At every show, try to make the judge aware of the effort that has gone into producing your work of art.

Honey jar label and other design classes

Read the schedule very carefully for this class. Make sure you comply with the current regulations: the schedule may sometimes rule on the content of the label.

The three main items marked on for this class are:
- Conformity to schedule or current regulations.
- Quality and clarity of the printing.
- Finally, and most importantly, the graphic design of the label.
- Non-conformity under the first point will get you disqualified, and the quality of the second point will go towards winning. The decider is always the graphic design.

Essays

The essay-writing classes have developed over the years: a very common class is the article for an association journal.

The schedule normally allows for the essay to be printed by the judge if they so wish, and for submission by email, enabling paper saving. Always conform rigorously to the word count: judges will have a number of essays to read and the last thing they require is an overlong read. Essays are judged on the following points:
- Always try to come up with an original point of view not commonly put forward.
- Being controversial will get your essay remembered.
- Lay out the argument with creativity.
- Good grammar
- Easy read, putting points over with a good writing style so that they are remembered
- Word count
- Read the schedule carefully and conform to its method of submission.
- Always table any references used with source detail, enabling the judge to verify.

Videos

Videos are really a modern extension of a written essay. Before getting into the finer judging points, make sure the device is on a stable platform and is in focus, that there is nothing in the background that you do not want shown and may distract from the point that the video is making.

- Stick to the time demanded by the schedule.
- Cover the subject required, with inspiration.
- Express the theme with good visual appeal and artistic merit.
- Use a uniqueness of presentation to enable the video to be remembered.
- Make sure that at all times you talk to the camera and not to the subject or hive that you are discussing.

5

Judging

The following notes are really for prospective judges, e.g. if you are not qualified as a honey judge, or are a keen exhibitor asked to judge a small show, and explain methods similar to those often used. There are many other methods so this should be read as a guideline.

As a judge you have a very responsible position in that exhibitors, who have gone to a lot of time and trouble, have the right to expect you to judge fairly and consistently. Also be aware that there are people who will try to bend rules and virtually cheat: it does not happen often, but be aware that it can. Always check if your steward is showing. If you are judging classes where your steward is exhibiting, make them aware of what you expect of them: which is to tell you in which classes they are showing and keep quiet until you have finished judging the particular class.

There are a number of marking sheets in this section. Feel free to copy and adapt to suit your own criteria: the marking sheets are there as a guide only. Please always use marking sheets when they are needed. There are many different marking sheets available to suit different judges, show managers, people, customs and processes.

I make no apology for repeating the following requirements for judging the classes: it is amazing how many prospective judges are not equipped correctly or don't use the PPE (personal protective equipment) when required.

Judges' tool kit and personal protective equipment (PPE)

- Wet cloths
- Lab coat
- Food industry hat, trilby or similar
- Disposable gloves
- Dry cloths
- Matches, or gas lighter
- Outside calliper
- Roofer's square
- Steel ruler
- Tape measure
- Large ham knife
- Refractometer (see instrument notes below)
- Hydrometer (see instrument notes below)
- Test weight traceable to NPL (National Physics Laboratory) standards: see explanatory notes below
- Scales (see notes on precision, accuracy and resolution)
- Spare batteries for the above scales
- Honey grading glasses must now be BD 2014 grade: purchase from Thorne's (see photo 46)
- Glass tasting rods (the Reid honey taster) or plastic disposables if you must. (See photo 47). Different sizes and end tools are used for different functions: liquid honey, soft set honey, heather honey, cut comb and frame for extraction.
- Plastic plate to enable judging of cut comb
- ISO wine glasses (see photo 48)
- Corkscrew
- Bottle opener (crown tops)
- Spittoon
- Apple or other items enabling you to clean your palate
- Torch
- Spare battery for the above torch
- Your own set of marking sheets
- BBKA list of plants that are worked by honey bees

If nucleus hives are to be judged: clean bee suit, tools and smoker – kit up as for a normal apiary inspection of your own hives. Make sure you have a strong plastic bag to seal up your suit and tools that you have used. Dispose of disposable gloves at the showground site. The whole bag contents can be disinfected and washed on arrival at home, keeping your own bees bio-secure and safe.

Instrument specification notes

In case of disputes with exhibitors, it is worth a judge's time to understand the terms and definition used for the relevant instrumentation used for judging exhibits.

Accuracy: is the closeness of the reading to the value of the item or exhibit being measured or is the standard being checked (test weight or standard solution)?

Precision and reproducibility: are the instruments able to give the same reading as the test standard repeatedly?

Resolution: what is the smallest part that can be read on the instrument? For example, a weigh scale could have a reading ability of 1 part in 2000 so a scale with a range of 2000g could read 1 gram. Or you could have a weigh scale that could read 1 part in 4000, so a scale with a range of 4000g would be able to read 1 gram.

So, to sum up, you could have an instrument with high resolution but poor accuracy and poor precision. An example would be if the instrument did not read the correct weight and was not consistent when rechecked with the same exhibit or standard. Your test weight should be traceable to the National Physics Laboratory standard.

The lesson is to always check your weigh scales in front of the show manager and steward before proceeding to judge and explain the different terms by demonstrating to them.

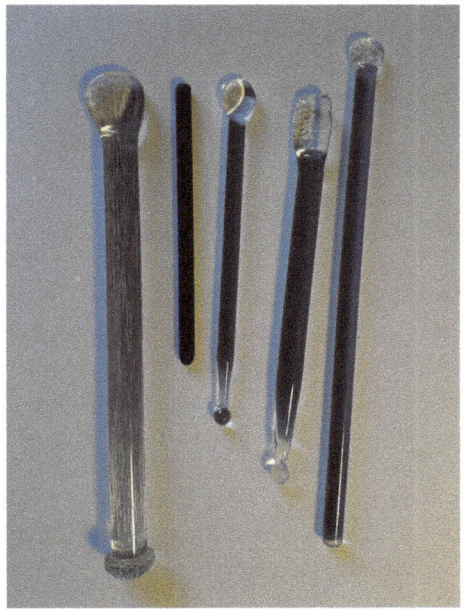

Photos, top to bottom:
46. Grading glasses (John Goodwin)
47. Glass tasting rods (John Goodwin)
48. ISO wine glasses (Gerry Collins)

Honey

Judging of extracted runny honey

Kit required

While judging, always wear a white laboratory coat and hat (a trilby or similar as used in the food industry). Make sure that your steward is similarly equipped.
- Refractometer
- Show schedule
- Set of BD grading glasses
- Torch
- Tasting rods: glass, or plastic if you must
- 2 honey jars or beakers used for both clean and used glass tasting rods
- Wet cloth
- Dry cloth

Method
- Do not partake for 24 hours before judging of any strong-tasting foods or drinks.
- It is advisable not to wear perfume or aftershave lotion, as this can affect the judgment of aroma.
- Beware of using lipstick or lip balm when judging.
- Check that the jar conforms to the schedule.
- Check that the lids conform: some shows still do not allow plastic lids.
- Firstly, check with the show manager that they have a set of grading glasses and how they differ in colour from your own. All judges should be using the BD glasses (named after Bernard Diaper, who the honey judging world has to thank for the work carried out in sourcing these filters and specifying them).
- Always use grading glasses against a matt white: your judge's steward lab coat is ideal. Do not use against bright or direct lights. (See photo 49).
- Starting with the light honey class, look for darkest on the bench and check it against the grading glasses; if darker than the light grading glass, put it out and repeat until the next darkest is lighter than the light grading glass filter.
- Check bottom of jars to make sure that the jars are of the same manufacture. Jars do not have to have the same mould number. Put out if they are different manufacturers. (See photo 50).
- Occasionally the two jars may look as though the honey is different. This can be caused by the light. Swap the two jars round to see if there is a difference, then hold both against your steward's coat.
- Check lids (do not open at this stage): they must match and be free of rust. For metal lids, check carefully that the plating matches and that there are no spots of rust on the lid turnovers. (See photo 51).
- Check outside of jars for cleanliness; it should be free from finger marks.
- Using your torch (see photo 52), check each jar for granulation, floaters, and foreign bodies. If there is any sign of these, put out the exhibit. If the jar is

Judging

49. Grading glasses (Gerry Collins)

50. Check the bottom of the jars (Gerry Collins)

51 (above). Check lids (Gerry Collins)

52 (right). Using a torch to check jars for granulation, floaters and foreign bodies (Gerry Collins)

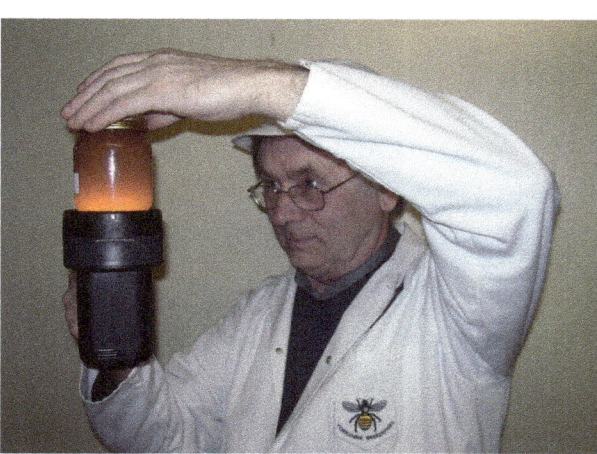

not full to the fill mark, again put it out: this is the only method you have of checking that the exhibit is to weight as you cannot weigh the contents of the jar without knowing the jar's tare weight. Use the torch to look for a light gap between the top of the honey and the bottom of the lid: this indicates an underfill. Any torch with a bright light is suitable. With the use of LED technology nowadays, you don't have to carry a big flashlight to get the necessary light energy. Modern LED torches also have the advantage of using rechargeable batteries.

- Open each jar: check for aroma by opening the jar and tilting the lid.
- Look very carefully on the bottom of the lid. Do not mark down for travel splash (fresh honey on the top of the lid).
- Check the lid threads for rust and black-coloured honey. This is an indication of honey that has got onto the threads and not been cleaned. The blackness is honey that has oxidised or picked up dirt, dust or ash.
- Check the top of honey for debris, bubbles or scum: put out if there are any signs.
- Check top of jar for dust. This dust is usually the result of the exhibitor not cleaning the lid when preparing the exhibit. The dust is talc used as a mould-release agent on the lid jar seal.

- Use your tasting rods to check for viscosity. If still doubtful put the exhibit on one side to check with a honey refractometer. The refractometer is used to measure moisture content. High moisture content honeys have a low viscosity: any show honey in the Liquid Honey section should be less than 20% water /moisture content (heather honey being the exemption).
- If you are in doubt of the moisture content of an exhibit, analyse with a refractometer calibrated for the moisture content of honey (this is very important as most refractometers calibrated for moisture analysis are scaled to read the Brix scale which is the inverse of moisture content). It is always prudent to make sure of the calibration of all instrumentation before using it for analyses. The standard zero check with a honey refractometer is to use distilled water.
- Check fully opened lid for dirty threads and internal rusting.
- Bring forward to the bench all remaining exhibits.
- Check for taste and aroma. Use the second jar for this as well as the first jar: you will sometimes find that the honey in a jar is the same colour as its pair but has a different taste. If there is any difference in taste, put it out. Move up and down the bench until you can place for prizes.

Judging soft set and granulated honey

Kit required
Always have on while judging a white laboratory coat and hat (a trilby or similar as used in the food industry). Make sure that your steward is similarly equipped.
- Magnifying glass
- Show schedule
- Torch
- Tasting rods, plastic or glass
- 2 honey jars or beakers
- Wet cloth
- Dry cloth

Method
- Do not partake for 24 hours before judging of any strong-tasting foods or drinks.
- It is advisable not to wear perfume or aftershave lotion as this can affect the judgment of aroma.
- Beware of using lipstick or lip balm when judging.
- Check that jar conforms to the schedule.
- Check that the lids conform: some shows still do not allow plastic lids.
- In summer at agricultural and floral shows, try and judge soft set honey classes as soon as possible due to the exhibits getting warm and starting to move (becoming liquid).
- Check lids (do not open at this stage). They must match and be free of rust. For metal lids, check carefully that the plating matches and there are no spots of rust on the lid turnovers.
- Check bottom of jars to make sure that the jars are of the same manufacture. Jars do not have to be same mould number. Put out if they are different manufactures.

53. Particles and bits at the bottom of the jar (Gerry Collins)

- With your magnifying glass, rotate the exhibit looking for any bits or particulate matter. Especially look on the underside of the jar as most particles and bits will gravitate to the bottom of the jar. If any particulate matter is found, put out the exhibit. (See photo 53).
- Check outside of jars for cleanliness.
- If the jar is not full to the fill mark, put out. This is the only method you have of checking that the exhibit is to weight as you cannot weigh the contents of the jar without knowing the tare weight. Use the torch to look for a light gap between the top of the honey and the bottom of the lid: this indicates an underfill. Any torch with a bright light is suitable. With use of LED technology, you don't have to carry a large flashlight to get the necessary light energy. Modern LED torches also have the advantage of using rechargeable batteries.
- Open each jar and check for aroma by just tilting the lid slightly. If you get the smell of beer on opening the jar, the honey is starting to ferment. Fermentation is quite often found in these classes. If the honey moves on tilting, put it out. Soft set honey should have a smooth, matt, dry surface with no movement.
- Check the lid threads for rust and black coloured honey. This is an indication of honey that has got onto the threads and not been cleaned. The blackness is honey that has oxidised and picked up dirt.

- Check top of honey for debris and bubbles or scum: put out if there are any signs.
- Check top of jar for dust. This dust is usually the result of the exhibitor not cleaning the lid when preparing the exhibit. The dust is talc used as a mould-release agent on the lid jar seal.
- Check fully opened lid for dirty threads and internal rusting.
- Bring forward to the bench all remaining exhibits.
- Check for taste and aroma. Use second jar for this as well as the first jar (you will sometimes find that the honey in a jar is the same colour as its pair but has a different taste. If there is any difference in taste, put it out.) Move up and down the bench until you can place for prizes.
- If you are judging granulated honey, you should expect to feel some granulation on the tongue. With soft set honey there should be no granulation, just a smooth texture as if the honey was runny. Soft set honey should not be very hard if prepared properly.

Judging heather honey

Kit required
- Always have on while judging a white laboratory coat and hat (a trilby or similar as used in the food industry). Make sure that your steward is similarly equipped.
- Show schedule
- Torch
- Tasting rods, plastic or glass
- 2 honey jars or beakers
- Wet cloth
- Dry cloth

Method
- Do not partake for 24 hours before judging of any strong-tasting foods or drinks.
- It is advisable not to wear perfume or aftershave lotion as this can affect the judgment of aroma.
- Beware of using lipstick or lip balm when judging.
- Check that the jar conforms to the schedule.
- Check that the lids conform: some shows still do not allow plastic lids.
- Check lids (do not open at this stage). They must match and be free of rust. For metal lids, check carefully that the plating matches and there are no spots of rust on the lid turnovers.
- Due to the thixotropic nature of heather honey, air bubbles are present. These bubbles should be evenly spaced in both exhibits. (See photo 54).
- Check bottom of jars to make sure that the jars are of the same manufacture. Jars do not have to be same mould number. Put out if they are different manufactures.

54. Air bubbles in honey (Gerry Collins)

- Check outside of jars for cleanliness.
- Use your torch to check that the jar is full. There used to be jars specifically made for heather honey with a bigger capacity (as heather honey will have bubbles). This is the only method you have of checking that the exhibit is to weight as you cannot weigh the contents of the jar without knowing the jar's tare weight. By using the torch you are looking for a light gap between the top of the honey and the bottom of the lid: this indicates an underfill. Any torch with a bright light is suitable. With use of LED technology, you don't have to carry a large flashlight to get the necessary light energy. Modern LED torches also have the advantage of using rechargeable batteries.
- Open each jar and check for aroma by opening the jar and tilting the lid. If the honey moves on tilting, put out.
- Use your tasting rod to create a groove on the surface of the honey. This will indicate the thixotropic nature of the honey. (See photo 55).

55. Creating a groove on the surface of the honey (Gerry Collins)

- Check the lid threads for rust and black-coloured honey. This is an indication of honey that has got onto the threads and not been cleaned. The blackness is honey that has oxidised and picked up dirt.
- Check top of honey for debris and bubbles or scum. Put out if there are any signs.
- Check top of jar for dust. This dust is usually the result of the exhibitor not cleaning the lid when preparing the exhibit. The dust is talc used as a mould release agent on the lid jar seal.
- Check fully-opened lid for dirty threads and internal rusting.
- Bring forward to the bench all remaining exhibits.
- Check for taste and aroma (use second jar for this). Move up and down the bench until you can place for prizes.

Judging the black jar class

This class is judged on aroma and taste alone. One of the problems that judges have with this class is that all honeys have to be tasted. However, if the judge feels that the honey is not of a clean sample, he or she is perfectly entitled to not taste the exhibit. This class is the most disliked by judges: sometimes if the judge is lucky the black jar class is judged by committee (either the public or association members).

Kit required
- Always have on while judging a white laboratory coat and hat (a trilby or similar as used in the food industry). Make sure that your steward is similarly equipped.
- Tasting rods
- 2 beakers or honey jars

Method
Line all the exhibits up in a line, start from one end of the row and check for aroma. Bring the best samples forward to the front of the row. Then check for taste if you are happy with the hygiene of the exhibit. By filtering out the best aromas first, this limits the amount of honey that has to be tasted, stopping the palate going into overload.

Judging the composite class

This class is a complex class to judge, if you do not have a lot of judging experience.

Kit required (depending on the complexity of the class):
- Always have on while judging a white laboratory coat and hat (a trilby or similar as used in the food industry). Make sure that your steward is similarly equipped.
- Show schedule
- Set of BD grading glasses
- Magnifying glass (there may be granulated and soft set honey in the exhibits)

- Torch
- Tasting rods
- 2 beakers or honey jars
- Marking sheets

The kit for this class could cover most of the judge's tool kit as you could have mead, wax, candle, cut comb and a frame for extraction as part of the class.

This class should always be judged using a marking sheet by summation of the results. The composite class can include a number of different exhibits. Some shows only schedule different types of honey; other shows and classes can include wax, mead, candle and confectionery. This is one class that can only have disqualification for a missing item. All items are judged and if they are of poor quality, they end up with a zero mark.

Method
- Do not partake for 24 hours before judging of any strong-tasting foods or drinks.
- It is advisable not to wear perfume or aftershave lotion as this can affect the judgment of aroma.
- Beware of using lipstick or lip balm when judging.
- Make sure prior to judging that all the necessary pieces per exhibit conform to the show schedule. If any exhibit has any pieces missing, inform the show manager before proceeding to judge
- Always use a marking sheet for a composite class.
- Get your steward to fill in the exhibitor numbers on the marking sheets.
- Check each item separately. The routine is:
  - Start with the mildest tasting exhibits first e.g. light honey from each exhibitor's entry. If judging a liquid honey, check with the grading glasses that the exhibitor conforms to the scheduled colour band.
  - If there is mead, a good plan is to do that first, especially if there is a dry mead.
  - Check lids (do not open at this stage): they must be free of rust and with no spots of rust on the lid turnovers.
  - Check outside of jars for cleanliness.
  - Use your torch to check that the jar is full to the fill mark. This is the only method you have of checking that the exhibit is to weight as you cannot weigh the contents of the jar without knowing the jar's tare weight. By using the torch you are looking for a light gap between the top of the honey and the bottom of the lid: this indicates an underfill. Any torch with a bright light is suitable. With use of LED technology, you don't have to carry a large flashlight to get the necessary light energy. Modern LED torches also have the advantage of using rechargeable batteries.
  - Open each exhibitor's light honey jar in turn and check for aroma by opening the jar and tilting the lid.
  - Check the lid threads for rust and black coloured honey. This is an indication of honey that has got onto the threads and not been cleaned.

The blackness is honey that has oxidised and picked up dirt. At any sign of rust, mark down.
- Check top of honey for debris and bubbles or scum. If there are any signs of debris, mark down.
- Check top of jar for dust. This dust is usually the result of the exhibitor not cleaning the lid when preparing the exhibit. The dust is talc used as a mould release agent on the lid jar seal.
- Check fully opened lid for dirty threads and internal rusting.
- Check for taste and mark accordingly.

- When all the above checks have been made, fill in the marking sheet for all the exhibitors' exhibits.
- Proceed to the next honey (i.e. the next mildest-tasting medium or similar, following the same routine).
- If there is a soft set or a granulated item in the class, it is advisable to judge them early in the judging cycle as they are very often very mild in taste.
- If there are any other honey show categories scheduled in the composite class, judge as explained under the relevant categories. The only difference is that no item or part of exhibit can be disqualified on judging parameters but they can be marked to zero.

Below is a typical marking sheet for the composite class (scoring range to suit judge).

Composite marking sheet							
Venue: Date: Class number:			Judge: Steward:				
Entry number	Item 1	Item 2	Item 3	Item 4	Item 5	Item 6	Total

All the above marks are just for guidance. These marks can be modified to suit the show manager's and judge's preferences.

Wax

Judging single block

Kit required
- Always have on while judging a white laboratory coat and hat (a trilby or similar as used in the food industry). Make sure that your steward is similarly equipped.
- Show schedule
- Scales (check weight): see note.
- Ruler
- Tape measure
- Dry cloth
- Torch

Method
- The aroma can and should be checked later in the procedure. If the ambient temperature is on the cool side, it is sometimes beneficial to rub the wax against a piece of cloth, care being taken not to scratch the piece.
- Check to see if the showpiece is level and/or symmetric (see photo 56).
- Check bottom of the piece for level and filling dip.
- Check for colour (it should be a shade of yellow similar to a primrose flower) and bring down your preference onto the bench.
- Both the container and the wax exhibit should be spotlessly clean: no hairs, lint, pieces of food, wildlife, bee parts or vermin. To check for any contamination within the block, a torch is shone through the wax exhibit, checking its translucence and for any internal contamination bits.
- Dimensions of the exhibits: check against the show schedule. Some shows specify thickness, diameter and shape of the wax piece. If any exhibit does not conform to schedule, it should be disqualified.
- Weight: use your scales. It is prudent to show the accuracy of the scale to the show manager (who should request that he/she is allowed to check the scale with a weight traceable to the NPL). If the show manager has not checked the scale, do not proceed with any weight checks of the exhibits until you have made sure your judge's steward is happy with your weigh scales' accuracy (which can be proven with your own check weight).
- The finish of the wax block should be free from any blemishes, scratch marks, cracks, chips and ripples.

56. Checking level and/or symmetry (Gerry Collins)

Judging small wax blocks 28g

Kit required
Always have on while judging a white laboratory coat and hat (a trilby or similar as used in the food industry).
Make sure that your steward is similarly equipped.
- Show schedule
- Ruler
- Tape measure
- Dry cloth
- Torch

57. Judging small wax blocks (Gerry Collins)

Method
Judging small blocks is similar to the above procedure, making sure that your scales are weighing accurately before proceeding with the following judging procedure (see photo 57):
- The first obvious check is to make sure that there is the correct number of pieces according to schedule.
- Check whether the schedule states total weight of blocks or individual weight.
- Check the weight of each individual piece.
- Each piece should be clean with no scratches or chips.
- Check to see if the showpieces are level and/or symmetrical.
- The bottom of the piece should be flat with no ripples.

- All pieces should be to within the weight tolerance and be of a similar weight.
- Make sure the correct moulds have been used according to the schedule.
- Wax pieces should be within a bee-proof container.

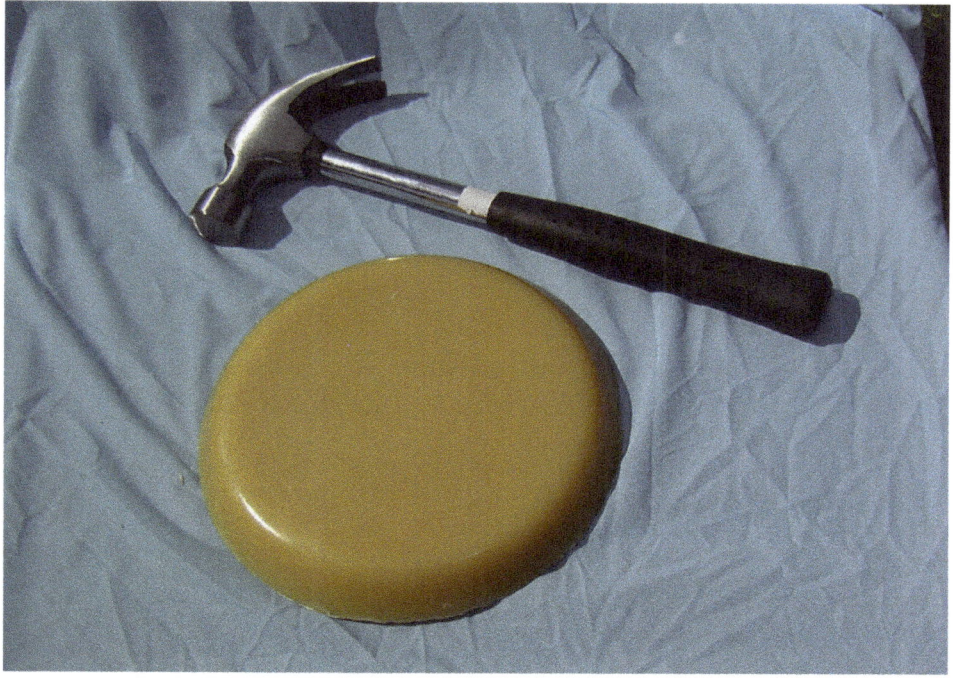

58. Judging commercial wax blocks (Gerry Collins)

Judging the commercial wax block

Kit required
(See photo 58).
- Always have on while judging a white laboratory coat and hat (a trilby or similar as used in the food industry). Make sure that your steward is similarly equipped.
- Show schedule
- Scales (check weight): see note.
- Ruler
- Torch
- Hammer

Method
Commercial wax blocks are judged to schedule, which will normally specify that the piece be judged on aroma, colour and crystal structure of the wax when broken.

The process is similar to the previous wax classes with a minimum weight normally scheduled.

There is sometimes also a requirement for a certain thickness.

Should be shown in a bee-proof container.
Check for aroma both before and after breaking.

Judging candles: various manufacturing methods

Kit required
Always have on while judging a white laboratory coat and hat (a trilby or similar as used in the food industry). Make sure that your steward is similarly equipped.
- Show schedule
- Ruler and/or joiner's or roofer's square
- Magnifying glass
- Marking sheets if not offered by the show management
- Calliper or some method to check diameter
- Method of igniting the candle: a simple box of matches or gas lighter

Candles should be judged using the summation method. The only exception to this would be if the exhibited candles do not conform to the scheduled dimensions (figurine-type candles).

You will find in the schedule at most shows that different methods of production are allowed. "Weighting" should to be given to the method of production:
- Poured (hardest)
- Dipped
- Cast in glass moulds
- Cast in silicone or similar moulds
- Rolled from foundation but allow extra if you find the foundation is home-made.

The candles will be made of pure beeswax and prepared as stated above: moulded, poured, dipped or rolled in accordance with the show schedule.

It is important that as a judge you understand the consequence of a wrong-sized wick for the diameter of the candle being shown.

Judging is by summation. Below is a suggested marking sheet, which does not include a loading for production method.

Method
At each stage of the judging, get the steward to fill in the marking sheets.
- Check that the show management have no marking sheets of their own issue before using your own marking sheets.
- Check length dimensions by laying against a tape-measure or ruler. This check is important with very long candles to ensure they conform to schedule (length does not include the wick). If judging a pair or trio, ascertain that that they are all to the same length. (See photo 59).
- Check diameter against schedule and use an outside diameter calliper: either electronic or old-fashioned mechanical one. Check the mechanical calliper diameter result with a ruler. (See photo 60 by John Goodwin.)

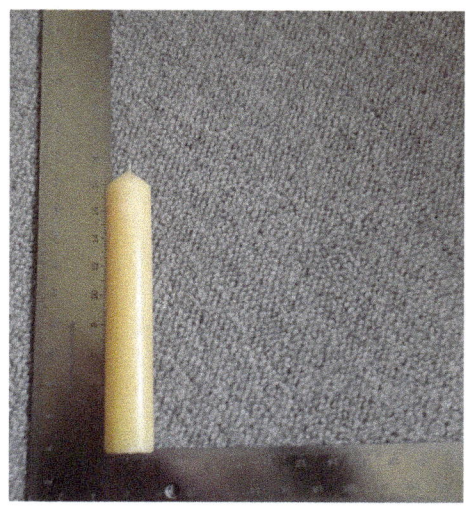
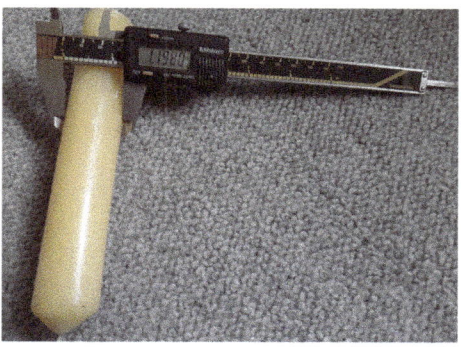

59 (left). Measuring a candle (John Goodwin)
60 (above). Mechanical caliper (John Goodwin)

- Smell the aroma.
- Check the bottom of the candle for centralised wick.
- Check for damage e.g. chipped edges.
- Look for any debris/impurities.
- The judge will light one candle to check its burning (indicating the correct sizing of the wick), unless the show schedule specifically requires him/her not to do so. The candle is allowed to burn for a while to enable to check for guttering and tunnelling (correct wick sizing for the diameter of the candle).
- The candles are now checked for smoking. Smoking indicates one of two things, one being impurities in the wax. Pure beeswax always burns clean, hence beeswax candles are used in churches and were used by the nobility in times past. Smoking can also indicate poor combustion of the wax, normally caused by wrong wick size.

The candle is relit after a time to allow the candle to cool, to make sure there is still wick available for re-ignition.

Candle marking sheet									
Venue:				Judge:					
Date:				Steward					
Class number:			Entry number						
Criteria	Max. Marks								
Wax; Colour /aroma	3								
Wax; hygiene	3								
Wick; centred, waxed and clean	3								
Finish; Cut base no chips	5								
Dimensions; same for all candles diameter and length	3								
Proportion; height to diameter ratio	3								
Burning; lighting guttering tunnelling	5								
Burning; Re-light, guttering, wick	5								
Total	30								
Award									

All the above marks are just for guidance. They can be modified to suit the show manager's and judge's preferences.

Wax models and flowers

This class is normally judged on the artistic interpretation of the schedule, the hygiene of the wax and the model's realism. It should always be marked by summation; the only disqualification would be non-conformation to schedule, including piece number, size of display, mechanics and props. (An example would be the schedule asking for fruits and the exhibitor showing vegetables.)

Wax model marking sheet										
Venue:				Judge:						
Date:				Steward						
Class number:			Entry number							
Criteria	Max. marks									
Wax; Colour /aroma (optional)	10									
Wax; hygiene/ clean	10									
Finish;	10									
Dimension; use of allowable space (optional)	10									
Proportion;	10									
Artistic; interpretation	50									
Total	100									
Award										

All the above marks are just for guidance, and can be modified to suit the show manager's and judge's preferences.

Mead

If the show allows, try to judge mead the first of the classes that have to be tasted, before honey, and certainly before confectionery (a common mistake made is to leave the mead till last).

There is a lot of terminology used in describing mead, much of which is used to describe other alcoholic beverages such as wine and beer.

- Legs: this term is used to describe the clinging of the mead to the side of the glass when the liquid is moved away from the side. There is a lot of debate in wine circles about legs. The reason that legs are seen on the glass is the vaporisation of alcohol in the mead, leaving the water molecules behind on the glass. Obviously environmental conditions, both of the glass and of the ambient show facility (temperature and humidity), will also affect the adhesion of the liquid to the glass. Basically, adhesion of liquid to glass side indicates a chance of alcohol. Does it tell you anything about the mead? Well, it does tell that there is or was some alcohol in the glass and that it has evaporated at ambient room temperature. As for the quality of the wine, no, it does not tell us the quality of the exhibited mead.
- Balanced is a term used in relation to components of mead and wine: alcohol, acidity, tannin, sweetness and fruit concentration. All these components are in balance if one doesn't overpower the remaining components. An example would be a melomel mead: if it is too fruity, you can only taste the sweetness of the fruit and not detect the taste of alcohol, honey or any other balancing components.
- Clarity is how clearly you can see through wine or mead. There should be no cloudiness and no suspended matter. A wine with clarity has the reflective quality of glass crystal.
- Hazy wine is an indication of wine still working or that has been spoiled.
- Petillant is a description of bubbly wine or mead when it should be still, in the context of judging still mead. The mead is still fermenting or working.
- Bouquet is used to describe the aromas you detect from the wine, e.g. honey, fruit, spice.
- Body often used to describe the level of alcohol in the wine. Normally described as heavy, medium or light.
- As well as clearing the palate, it is a good idea to take time to allow your sense of smell to recover. A good technique is to inhale fresh filtered air through your lab coat sleeve.

Kit required
- Always have on while judging a white laboratory coat and hat (a trilby or similar as used in the food industry). Make sure that your steward is similarly equipped.

- Show schedule
- Marking sheets if not offered by the show management
- At least six ISO glasses if possible. Have as many glasses as exhibits for the class being judged.
- Spittoon (see photo 61).
- Bucket (to empty the content of the spittoon)
- Torch
- Corkscrew
- Wet cloth
- Dry cloth
- Palate clearers: whatever suits your own taste – beer, dry biscuits, cheese or fruit.

61. Spittoon (John Goodwin)

Method

Mead is always judged using a marking sheet: see below. Different judges have different marking sheets of their own which they have developed over the years. Some shows issue their own sheets and insist they should always be used. On completion of judging the classes, the totals are summated and the prizes awarded. The completed marking sheets should be signed by the judge and steward and then returned to the show secretary's desk. The judging process is as follows:

- Check that the show management have no marking sheets of their own issue before using your own marking sheets.
- Get your steward to fill in the exhibitor numbers on the marking sheets.
- Do not partake for 24 hours before judging of any strong-tasting foods or drinks.
- It is advisable not to wear perfume or aftershave lotion, as this can affect the judgment of aroma.
- Beware of using lipstick or lip balm.
- Dry mead should always be judged before the sweeter meads. As with all the other show products, cleanliness is important.
- Establish that all the entries conform to the bottles and corks scheduled.
- A good tip before you start to judge the mead: just look at all the exhibits in the class to be judged. Very often you will find an exhibit where the cork has started to push out of the bottle.
- Check for external hygiene of the bottles (fingermarks are quite common).
- Check for sediment: spin the bottle by placing your finger in the punt. If there is sediment, it will rise off the bottom of the bottle and floaters of varying pedigree and source will appear. It is not unknown to see hairs both from humans and various species of animals. These hairs are not seen until the bottle is spun due to the refractive index.

- If in doubt about the hygiene of the bottle's contents prior to opening, check with torch and mark down if you have any concern about the exhibit's hygiene and chemistry. Do not taste.
- Remove cork, check for hygiene and mark.
- Pour into an ISO glass, check for clarity and rotate to see if there are legs.
- Check the bouquet: never ever try and judge bouquet by inhaling from the neck of the bottle as you may pick up cork mould smell. If you are judging on a cold morning in a marquee at an agricultural show, be careful to allow the mead and the glass to come up to a reasonable temperature.
- In-between sampling the exhibits, take time to clear your palate with dry biscuits or fruit and cheese: whatever you feel achieves this.
- A good method for judging the mead is to line up all the exhibits and place your glasses before them again in a line parallel with the mead bottles.
  - Starting at one end, check for bottle hygiene and mark.
  - Check for colour and clarity and mark.
  - Spin the bottle, check for sediment and mark.
  - Open the bottle, check for hygiene of cork.
  - Pour into glass in front of the bottle.
  - Move down the row of bottles, repeating the above steps.
- When all the bottles have been judged,
  - Check for aroma of each sample.
  - Check for taste, balance, flavour and body.
  - Repeat up the row.
- Total up the marks. If the steward has done the marking, check the summation and sign off.

Judging

Mead marking sheet										
Venue:			Judge:							
Date:			Steward							
Class/type:		Entry number								
Class number:										
Criteria	Max. marks									
Presentation: bottle hygiene	5									
Presentation: Cork, hygiene	5									
Sedimentation;	10									
Clarity/colour;	10									
Bouquet;	10									
Flavour;	20									
Balance	20									
Body	20									
Total	100									
Award										
Remarks										

Judging honey beer

Kit required
- Always have on while judging a white laboratory coat and hat (a trilby or similar as used in the food industry). Make sure that your steward is similarly equipped.
- Show schedule
- At least six ISO glasses
- Spittoon
- Bucket
- Torch
- Corkscrew and crown bottle opener
- Wet cloth
- Dry cloth
- Palate clearers: whatever suits your own taste, e.g. sparkling water, dry biscuits, cheese fruit.

Method
- Do not partake for 24 hours before judging of any strong-tasting foods or drinks.
- It is advisable not to wear perfume or aftershave lotion as this can affect the judgment of aroma.
- Beware of wearing lipstick or lip balm.
- Line up all your exhibits and get your steward to fill in the entry numbers on the honey beer marking sheet.
- Only open one bottle at a time and go through the full process of judging.
- A good idea is to use ISO wine-tasting glasses. These glasses keep the aroma in the glass and also limit the size of the sample you take. Beware and only drink small amounts.
- In-between sampling the exhibits, take time to clear your palate.
- As well as clearing the palate, it is a good idea to take time to allow your sense of smell to recover. A good technique is to inhale fresh filtered air through your lab coat sleeve.
- As each exhibit is sampled, make sure your steward marks up the honey beer marking sheet.
    - Check the bottle for sediment. If there is sediment, treat the exhibit with care when pouring.
    - Pour into glass in front of the bottle.
    - Check clarity and mark.
    - Check for colour and mark.
    - Check for aroma of each sample.
    - Check for balance, flavour, body.
- When all bottles have been judged,
    - Take up glass and check bouquet.
    - Taste for balance.
    - Final mark for flavour.
    - Total up the marks. If the steward has done the marking, check the summation and sign off.

Beer marking sheet										
Venue:			Judge:							
Date:			Steward:							
Class/type:		Entry number								
Class number:										
Criteria	Max. marks									
Presentation; bottle, top, label	5									
Clarity; hazy, cloudy, dull, opaque, bright, crystal	10									
Colour; honey, caramel, brown, burnt malt,	5									
Bouquet; malty, dried fruit, honey, acidic	10									
Flavour; hoppy, Bitter, sweet, fruity, honey, toffee	50									
Balance; bitter, mild, porter	10									
Head; white, delicate, flat, frothy	10									
Total	100									

Judging confectionery

When judging the confectionery classes, check which classes have been allocated to you. Mark the classes up in the schedule: see which class you perceive to be the sweetest or with a lingering taste (chocolate being an example). If allowed, try to judge these classes last. It is a good idea to judge the confectionery section as a whole anyway, and always last of all the show categories.

Be very aware when judging the confectionery classes that you are judging some person's food. Always make sure that every effort is used to keep the food from being contaminated and always use disposable gloves. If you cut into a cake that is undercooked, make sure that you wash your knife thoroughly before proceeding to the next exhibit.

Honey cake or honey fruit cake

Kit required
- Always have on while judging a white laboratory coat and hat (a trilby or similar as used in the food industry). Make sure that your steward is similarly equipped.
- Show schedule.
- Ham knife, or similar (see notes on show management). Knives at some shows are provided by the show management, as knives are not allowed to be carried onto the show ground for security reasons.
- Disposable plastic gloves (make sure you wear them when touching the exhibits).
- Tape measure.
- Ruler.

Method
- Put on your plastic gloves and make sure your steward is also suitably attired.
- First make a visual check that the cake has been cooked in the correct size of baking tin. Disqualify any that do not conform.
- All other checks are markable and not disqualifiable except for the last point.
- Check before the cakes are removed from the container that the numbering is also on the plate so that the bag/container/dome can be removed.
- Check out appearance: mark down for smiley face (broken crust).
- Cut cake in half, open and check aroma.
- Make sure fully baked and not burned.
- Taste if you are happy the cake has been cooked completely (if in doubt do not taste). There should be a subtle taste of honey in most honey cake recipes scheduled for honey shows.
- Texture should have a residual crumb, and not be so moist and fine that the texture is like clay.

- Check to make sure it conforms to the schedule: correct fruit etc. Be careful! Some items of fruit vary in appearance. Red glacé cherries have been bleached in sulphur dioxide and then dyed. Organic cherries have much more subdued colouring and can look like sultanas or raisins when cooked.

As you are going through the above test, bring the better exhibits down to the front and line up your preference, checking against the marking sheets.

Total up the marks. If the steward has done the marking, check the summation and sign off.

Honey baking marking sheet										
Venue:				Judge:						
Date:				Steward						
Class number:		Entry number								
Criteria	Maximum marks									
External appearance;	10									
Internal condition;	10									
Aroma;	20									
Flavour;	20									
Texture;	20									
Total	80									
Award										

Hives

Judging the observation hive

In judging the observation hive, the most important aspects are not only the show quality but the welfare of the bees, and the safety of the public in the event of an accident. If you are not happy with any welfare issues, make the show officials aware of your concerns.

Welfare problems found are:
- Not enough bees in the hive, leading to bees being chilled.
- No PPE available in the case of an accident (check gaffer tape, bee suits and dust sheets are available. Yes, it is your responsibility.)
- No water or syrup spray available.
- Leaking of bees out of the observation hives
- Bees in sunlight leading to gross overheating. It has been known for there to be complete colony loss due to this problem.
- Obvious signs of diseases.

Kit required
- Always have on while judging a white laboratory coat and hat (trilby or similar as used in the food industry). Make sure that your steward is similarly equipped.
- Magnifying glass
- Show schedule

Method
Start your judging by reading the schedule very carefully. Some shows and associations go to a lot of trouble to write the schedule for this class, as their objective is to educate the public, so allow marks for good information props. A logical way to judge the observation hive is to work through a marking sheet as shown below. This marking sheet, like all the marking sheets supplied, is really only a guide to custom, and practices do change depending on the geographical area in which you are judging, so adaptations of the marking sheets are sometimes necessary.

If any reportable disease is found in the colony, stop judging the class. Inform the show officers of your findings. Make sure that the local bee inspector has been informed and get confirmation from the show officers that this action has been taken and what other action is being taken. Immediately bag up all your equipment ready to be washed and sterilised.

Judging

Observation hive marking sheet										
Venue:		Class number:			Judge:					
Date:		Entry number			Steward					
Criteria	Max. marks									
Feeder;	5									
Water spray;	5									
Bee space;	5									
Ventilation;	5									
Construction;	5									
Frame hygiene;	5									
Frame spacing;	5									
Drawn combs;	5									
Population, correct number of bees;	15									
Colony drones	5									
Workers;	5									
Pollen	5									
Stores;	5									
Queen marked;	5									
Hygiene;	5									
Propolis;	5									
Correct labelling;	10									
Foul brood if found disqualify;										
Chalk brood;	-10									
Sac brood;	-10									
Varroa;	-20									
Total	100									

Judging the nucleus hive

Kit required
- Bee suit
- Hive tools
- Plastic gloves (disposable)
- Boots or wellies
- Plastic bag, to pack all your kit including wellie boots and to be sealed for the return journey home.

Method
- Before starting to judge the nucleus hives, make sure that you are properly suited and that your steward is also suited and gloved.
- Check that the bees are in a cage so that there is no direct flying of bees into the public at head height and that there is little danger to the public.
- If you are not happy with any of the arrangements, make the show officers aware of your concerns before proceeding.
- If on opening a hive you find that the bees are very tetchy and of poor temperament, close the hive immediately.
- If any reportable disease is found in the colony, close up the hive immediately and cease judging the class. Inform the show officers of your findings. Make sure that the local bee inspector has been informed and get confirmation from the show officers that this action has been taken, and of what other action is being taken. Immediately bag up all your equipment ready to be washed and sterilised. Make sure all bee suits worn by the stewards are treated and bagged up the same.
- If all is well, carry on judging, one hive at a time, and close up each hive before proceeding to the next hive.
- Use a copy of the marking sheet attached or similar, and check each hive for the following:
  - Hive quality: woodwork rot and clean
  - Bee space
  - Frame spacing
  - Construction
  - Frame quality (even drawing)
  - Clean frames
  - Drawn combs
  - General colony balance
  - Colony drones
  - Workers
  - Stores
  - Pollen
  - Queen marked
  - Temperament

- Queen marked correct year
- Propolis
- Colony hygiene
- Foul brood: if found disqualified
- Chalk brood
- Sac brood
- Varroa

Get your steward to fill in the marking sheet. There is an example of a marking sheet below:

Nucleus hive marking sheet										
Venue:				Judge:						
Date:				Steward						
Class number:			Entry number							
Criteria	Max. marks									
Hive quality;	5									
Frame spacing;	5									
Bee space;	5									
Frame quality; even drawing;	5									
Construction;	5									
Frame hygiene;	5									
Frame spacing;	5									
Drawn combs;	5									
General colony balance;	15									
Colony drones	5									
Workers;	5									

Temperament;									
Pollen;	5								
Stores;	5								
Queen marked;	5								
Hygiene;	5								
Propolis;	5								
Correct labelling;	10								
Foul brood if found disqualify;									
Chalk brood;	-10								
Sac brood;	-10								
Varroa;	-20								
Total	100								

Judging a display

- Start by getting your steward to fill in the marking sheet with the exhibitor's number.
- Never attempt to judge something as complex as a display without a marking sheet.
- So that the ambiance, balance and set-up of the display is not disturbed, mark first for neatness and artistic merit, prior to all other criteria.
- Look for hygiene of wax.
- Quality of interpretation of wax models: mark accordingly.
- Mark the mead as you would a Mead class. Start with the light-coloured mead first as this is more likely to be dry.
- Make sure to open all the honey jars and check as you would for the individual classes in the Extracted Honey class.
- Check all the cut comb and sections as you would the individual Cut Comb and Section classes.
- Check the frame for extraction as you would for the Frame for Extraction class. Examine especially for vermin, wax moths, etc.

Display marking sheet										
Venue:				Judge:						
Date:				Steward						
Class number:			Entry number							
Criteria	Max. marks									
Neatness;	10									
Wax;	10									
Mead;	10									
Extracted honey;	10									
Dimensions; to schedule	10									
Comb honey;	10									
Artistic;	30									
Symmetry:	10									
Total	100									
Award										

Judging skeps

There are not many shows where skeps are exhibited nowadays and they are not a category included in the BBKA Judge Certificate.

Kit required
- Always have on while judging a white laboratory coat and hat (a trilby or similar as used in the food industry). Make sure that your steward is similarly equipped.

- Show schedule.
- Ruler, possibly an outside calliper.
- Tape measure.
- Torch.

Judging method that could be used

Check that the skep meets all the requirements of the schedule.
- Material of manufacture: be aware that in some areas of the UK other materials are used for skeps (rushes, willow, heather).
- No sign of mould.
- The skep is new and has not been used for handling or keeping of bees.
- Check that the skep is suitable for use as required by the schedule. Is it for catching swarms or keeping bees?
- Size.
- Turn skep upside down and shine a torch through it to check that it's bee-proof.
- Is the skep symmetrical?
- Check that the lapping is neatly spaced and of a space ratio to give a good binding to suit the rope diameter.
- Component parts that may be required in the schedule:
  - Base
  - Hackle
  - Cap (super)

Obviously the skep should be free-standing and not wobbly.

Handicrafts

This title covers a lot of different skills.

Kit required
- Always have on while judging a white laboratory coat and hat (trilby or similar as used in the food industry). Make sure that your steward is similarly equipped.
- Show schedule
- Ruler tape measure
- Magnifying glass

Judging method
- Check all exhibits for schedule conformity: dimensions, art form and piece number (some schedules allow for pairs).
- Divide up the exhibits into the different art forms.
- Mark on finish, skill required, and time to produce.

Children's class

If there was ever a class that gets judges into hot water, this is the one. Hopefully the show management have had the good sense to have a number of classes, covering as many age groups as possible. The only advice for this class is to give as many prizes as you are allowed. The really good exhibits should be celebrated notwithstanding by giving Firsts and Seconds but try and persuade the show manager to allow you to go down as far as commended which gives six cards per class: First, Second, Third, Very Highly Commended, Highly Commended and Commended.

Microscope slides

There are normally two main subject sections for the class: pollen and honey bee histopathology/anatomy. There is sometimes a class for photomicrographs which bridges the Microscope Slide and the Photograph classes.

Often two slides are requested by the schedule, and are normally judged on the quality of the mounting and the use of suitable mountant.

Kit required
- Always have on while judging a white laboratory coat and hat (a trilby or similar as used in the food industry). Make sure that your steward is similarly equipped.
- Show schedule
- Compound microscope, preferably trinocular with a monitor with a decent substage and Abbe condenser (this item should be provided by the show management. If you are not happy with the quality of microscope supplied to judge this class, do not proceed.)
- If you are not happy or do not have the relevant experience (which is the best bit of kit you can have for this class) to judge this class, please make this known to the show manager.

Photographs

Kit required
- Always have on while judging a white laboratory coat and hat (a trilby or similar as used in the food industry). Make sure that your steward is similarly equipped.
- Ruler
- Tape measure

Method

Care needs to be taken in this section: if the schedule is not specifically *Apis mellifera*, other bee species are allowed as the subject matter. The problem can arise if the actual insect photograph is not a bee. It has been known for awards to be given for hoverflies. So, make sure you are good at species recognition. The most common photograph shown is a queen bee being attended to by her workers. For some reason beekeepers think this makes for a wonderful subject, which of course it is but not very original as a photograph. Of all the different categories that are part of a honey show, photography is the one category that allows novice beekeepers to show and be as successful as the more experienced beekeeper/exhibitor.

Make sure all exhibits conform dimensionally to the schedule and disqualify if not.

Mark for:
- clarity and focus
- originality
- composition and design
- adherence of subject matter to the schedule
- storytelling, humour
- innovation
- colour coordination and style

Videos

If you are asked to judge videos or a class of a similar medium, it is very likely to be judged prior to the show at your premises or other premises than the show ground.

Videos are really a modern extension of a written essay. Before getting into the finer judging points, if background music is used, make sure that there is a certificate of authorisation to play the said music.

- Make sure video is to the time specified in the schedule.
- Make sure the theme required is covered.
- Notice how much time the presenter spends looking at the camera explaining the subject.
- How professional is it? The overall impression of the workmanship.
- Do you remember the points made?
- If a controversial point of view is put over, how good are the arguments?
- Is the subject covered with inspiration?
- Is there good visual appeal and artistic merit?

Open judging

Open judging is not normally expected of a novice judge. The show management should allow at least a third more time than what would be normally allowed for the said number of exhibits. Open judging is carried out in two theatres or situations:

- Firstly, at a small association or branch level where the organisation uses open judging for the benefit of educating their members in what is required of their exhibits for showing. The results of open judging in this environment are quite outstanding. There have been many instances where exhibitors at small show level have improved the quality of their exhibits and their confidence to go on to compete and take awards at County or National Honey Show level.
- Secondly, some of the larger agriculture and floral shows have traditionally judged in front of the paying public. It is perceived by the show management that this is all part of the ethos of rural showing. If you are judging in this environment, carefully brief your steward to make sure that every time you are stopped by the public, he or she knows exactly where you are up to in the judging of the exhibits. The public may stop you judging and ask questions which to them are very pertinent and important, and to you at best a distraction. One idea is to give a little series of presentations: in this way you control what stoppages you are prepared to allow. A good example of this when judging the mead is to explain what mead is and what you are looking for in a show exhibit.

A lot of patience should be shown on the questions asked. Some of the public can be very interesting: a few years ago some very old beekeeping memorabilia, historically important to the local association, was given to the judge to pass on to the local beekeeping officers. Honey shows have been held in the middle of shopping malls or garden centres, just to raise the profile of beekeeping both generally and in the local environment.

Final speech

Very often as a judge you are asked to comment on the quality of exhibits and also give advice. Before any comments are made, it is good manners to ask for a round of applause for your steward or stewards who have given their time and also possibly given up an opportunity to attend some other function during the judging period. Secondly, always thank your hosts and the show manager for inviting you to judge.

Your comments on the exhibits should always be constructive and give encouragement. There is nothing worse than a judge who takes satisfaction in criticising for the sake of it. Start by explaining that some exhibits did not even get to the stage of being tasted, and explain why. On no account pick an exhibit as an example. Try and pick out really good examples, and encourage the exhibitor to exhibit at the next higher level – be it the local county association or the National Honey Show. Also spend some time talking about the novice exhibits and give lots of encouragement. It is very easy to criticise and come across as a know-all prat but it takes a person with intellect to praise. By giving encouragement you will help increase the size and quality of honey exhibiting nationally.

About the author

John Goodwin has kept bees in his home county of Cheshire for about thirty years, usually about forty colonies, and always focusing on producing honey from varying sources. The honey forage crop in Cheshire is from five main geographical environments: semi-urban areas, pastoral farmland for dairy farming (which gives a good crop of clover honey), a relatively small acreage of arable land producing oil seed rape and borage, and an even smaller acreage of game cover with phacelia and mixes of buckwheat. Finally, and still within Cheshire, the Peak District heather moors, conserved for grouse shooting, produce high volumes of very good quality heather honey. There are not many years when a good variety of show quality honey cannot be produced within Cheshire.

As Honey Show Manager for a number of years, both of the Cheshire Beekeepers Association Honey Show and Nantwich International Cheese Show Honey and Bees section, John has always tried to facilitate the training of and assessments by future honey judges, making sure that time was allowed for Honey Show judge assessments and classes not normally scheduled in the majority of honey shows (such as microscope slides, observation hives and displays). He encourages novice beekeepers from all backgrounds to exhibit, making the show inclusive and available for all beekeepers, and encouraging novice beekeepers to show at major national honey shows.

John has held the BBKA Judge certificate for well over 15 years and has been a Judge Certificate Assessor for over 10 years, judging honey all over the UK, usually up to twenty shows per year. He has recently undertaken prospective Honey Show Judge Assessments and final Judge Assessments at the National Honey Show.

He has given workshops on showing, show judging, honey tasting, and producing show honey and wax at the National Honey Show and the BBKA Spring Conventions, BBKA associations and branches all over the UK.

About the author